The World According to Military Targeting

Prisms: Humanities and War

Edited by Anders Engberg-Pedersen

War and Aesthetics: Art, Technology, and the Futures of Warfare, edited by Jens Bjering, Anders Engberg-Pedersen, Solveig Gade, and Christine Strandmose Toft

The World According to Military Targeting, by Erik Reichborn-Kjennerud

The World According to Military Targeting

Erik Reichborn-Kjennerud

The MIT Press
Cambridge, Massachusetts
London, England

The MIT Press
Massachusetts Institute of Technology
77 Massachusetts Avenue, Cambridge, MA 02139
mitpress.mit.edu

© 2025 Massachusetts Institute of Technology

This work is subject to a Creative Commons CC-BY-NC-ND license.

This license applies only to the work in full and not to any components included with permission. Subject to such license, all rights are reserved. No part of this book may be used to train artificial intelligence systems without permission in writing from the MIT Press.

The MIT Press would like to thank the anonymous peer reviewers who provided comments on drafts of this book. The generous work of academic experts is essential for establishing the authority and quality of our publications. We acknowledge with gratitude the contributions of these otherwise uncredited readers.

This book was set in Deca Serif by Westchester Publishing Services. Printed and bound in the United States of America.

Library of Congress Cataloging-in-Publication Data

Names: Reichborn-Kjennerud, Erik, author.
Title: The world according to military targeting / Erik
  Reichborn-Kjennerud.
Description: Cambridge, Massachusetts : The MIT Press, 2025. |
  Series: Prisms : humanities and war | Includes bibliographical references
  and index.
Identifiers: LCCN 2024035310 (print) | LCCN 2024035311 (ebook) |
  ISBN 9780262552349 (paperback) | ISBN 9780262383158 (pdf) |
  ISBN 9780262383165 (epub)
Subjects: LCSH: Strategy. | Military planning. | Geopolitics. | Balance of
  power.
Classification: LCC U150 .R44 2025 (print) | LCC U150 (ebook) |
  DDC 355.4—dc23/eng/20241129
LC record available at https://lccn.loc.gov/2024035310
LC ebook record available at https://lccn.loc.gov/2024035311
10 9 8 7 6 5 4 3 2 1

EU product safety and compliance information contact is: mitp-eu-gpsr@mit.edu

Contents

Acknowledgments   vii

1   Welcome to the World of Military Targeting   1

2   Military Targeting as Epistemology   25

3   The Enemy as a System: The Geopolitics of Strategic Paralysis   47

4   Hamlets, Humans, and Worms: Computing the Environment in Vietnam   87

5   FEEDing the Network: Hunting for Signatures   129

6   Tinker, Tailor . . . Trial and Error: Experimental Design and the "Algorhythmic"   169

7   Conclusions, or Future Beginnings   209

Notes   223
Index   267

Acknowledgments

The process of writing this book has been an endeavor that would not have been possible without the people I encountered in the process. Although I have never had the chance to meet most of the individuals that are referenced across these many pages in person, their work, has nevertheless been fundamental for making this book possible. But there are also many people who I have encountered and who have been of tremendous help over the years in different capacities.

Besides my gratitude to the many people who have trotted along the hallways at NUPI over the years, in IT, the library, and the others who make the wheels go around, several individuals at NUPI deserve individual expressions of appreciation: Iver Neumann for spurring me on to undertake this great ordeal and for the many conversations that started to fill the somewhat blank canvas; Karsten Friis for relentlessly challenging and encouraging me; Ole Jacob Sending for invaluable guidance and insightful remarks on many drafts along the way; and Morten Skumsrud Andersen for his very helpful comments and suggestions on patchy fragments, to which I hope I have made justice. There are many more at NUPI whose names would be too numerous to mention, but who have commented on early drafts with excitement and with whom I have been able to share frustrations.

Outside the confines of the institution, I would especially like to thank Tim Stevens for his enthusiasm for the project along with great feedback in its embryonic stages and to Christian Bueger for his thorough reading and very challenging and insightful remarks and recommendations on an early draft.

A very heartful thank you goes Antoine Bousquet and Kyle Grayson for their incredible persistence in reading the whole manuscript and providing great feedback. Special gratitude also goes to Aggie Hirst for her detailed reading, very insightful comments, and suggestion, but not the least for all her words of support.

Without you, Claudia Aradau, this project would have been stranded a long time ago. Thank you for all your help with everything. Your faith in the project, tireless efforts, inspiration, and generosity have been invaluable.

At MIT I would like to extend my sincere gratitude to Anders Engberg-Pedersen, whose belief in the project and very helpful guidance have made this possible. And to Victoria and Gabriela, thank you for your persistence and all your help.

I owe a special debt to my mum and dad for encouraging me to be curious about the world and inspiring me to seek out its many mysteries. Last, but not least, I am endlessly thankful for my little family: Frida for your extraordinary patience and for bearing with me (which is not a small feat) and Irlin and Ludvig for bringing enormous joy and meaning into an all too often incongruous world.

1

Welcome to the World of Military Targeting

"Track 'em and Whack 'em"

> We had struck pay dirt, almost haphazardly, early on in Afghanistan, killing an al-Qaeda leader and several of his lieutenants as they were fleeing Kabul. The strike was enabled by combining imagery and intercepts in real time. After I was briefed on the operation, I asked, "Why can't we do that all the time?" and put some bright minds on figuring out how to institutionalize the approach. We set up an effort, the Geocell, staffed it with smart young folks, teamed them up with imagery analysts from the National Geospatial-Intelligence Agency (NGA), and then wired them directly to tactical units in the field.... When visitors passed through the cypher-locked door and entered the secure work area, though, they could see huge screens with current imagery and watch Fort Meade analysts in multiple chat rooms throughout the war zone, advising combat forces in real time and living up to their self-described role: "We track 'em, you whack 'em.[1]

This mise-en-scène from the memoirs of Michael Hayden, former director of both the US National Security Agency (NSA) and the Central Intelligence Agency (CIA), describes how the American war-fighting machine was reconfigured and revamped for what was described by the US administration in the early days of the "war on terror" as a global manhunt.[2] It also reveals an ensemble of various normative and epistemological assumptions, techniques and technologies, everyday practices and institutions, operators, and targets that are illustrative of how the US military and its martial organs reorganized the way they imagined, assembled, and waged war.

Reimagining war as a manhunt against networked individuals—presumably elusive, hidden, and mobile—was a far cry from the monolithic threat of the Soviet Union with its fixed infrastructural systems that provided the US military with an abundance of potential targets to strike. The key challenge, as Hayden would later outline, was that during the Cold War "that enemy was pretty easy to find. Just hard to kill. This was different. This enemy was relatively easy to kill. He [sic] was just very, very hard to find."[3]

Seeing the "find/kill–easy/hard" calculus as inverted, new approaches were needed. Experimentally conjoining techniques, technologies, and practices from various military, intelligence, and security branches, the US war machinery had, as Hayden stated, "almost haphazardly" stumbled upon a novel way of waging war. This "pay dirt" found its form in a networked structure, later institutionalized as the Geocell—a heterotypic cell fusion linking the "eyes" of the NGA and the "ears" of the NSA[4] with operations on the ground. Rewired, retrained, and reorganized to respond to what the US military saw as the sanctuaries for modern insurgents, this proved to be a critical shift for the US war machinery.

While insurgents have always found refuge in complex terrains, be it natural or urban environments, the electromagnetic spectrum and digital communication networks were now perceived to provide new "sanctuaries." Not only could insurgents hide in complex terrains, but now, it was assumed, they could also communicate across time and space through digital networks, coordinate and self-synchronize, popping-up to strike and blend back into their safe havens. However, as insurgents moved their communications to digital networks, they were simultaneously self-documenting and storing their chatter, locations, and movements in digital form—a treasure trove for the martial apparatus.

This was to be the "golden age of signals intelligence."[5] The NSA was moving from passive signals intelligence (SIGINT)—where one hoped that transmissions would serendipitously enter their sensors—to active SIGINT activities—sweeping up massive amounts of metadata and digital break-and-enter operations to exploit digital communications around the globe. SIGINT had now expanded from the midpoint of a communication network to the end points of targeted networks.

On the other side of this cell fusion, the NGA was moving from printed maps dotted with photos and location pins toward digital geospatial intelligence (GEOINT). Using satellites, various radars, and other sensors, the NGA produced and visualized diverse digital 3D maps and images and layered these with detailed real-time data about every dot, from ground conditions to humidity in the air to buildings, individuals, and vehicles, visualizing the changing operational environment. The effect of experimentally configuring imagery and communication intercepts in real time and layering them onto digital maps was a radical reconstruction of the battlespace with the ability to track friends and foes through physical spaces. Wiring the heterotypic cell fusion onto and into this novel spatiotemporal model, Geocell's "tracking and whacking" operations were under way.

While the US military continues to highlight the presumably hidden, mobile, and networked enemy that served as the guiding imaginary for manhunts, the presumption that these individuals could simply be found, tracked, and killed would prove problematic. Not only did so-called insurgents and terrorists look like and behave like everyone else, but the problem was more fundamental. It concerned who and what the enemy is. But the enemy is not simply found "out there"; they must be "made up."[6]

The task was thus not only about how to reorganize the US military to distinguish between friend and foe and finding and killing individuals, but how to reconfigure the whole way in which the martial apparatus made the enemy known. To produce and operationalize this supposedly elusive enemy figure, the US military would mobilize and update a particular practice that had been utilized, tested, and battle-hardened over the past hundred years, namely, *military targeting*.

From its emergence in the interbellum years and throughout World War II and the Cold War, to the urban terrains and jungles of Vietnam, the mountains of Afghanistan and desert landscapes of Iraq, and to an imagined future conflict with China, military targeting has for over a century been fundamental to how militaries see the world, imagine war, produce their enemies, and assemble their forces and how they conduct warfare. Operating as a mode of inquiry into the world, abstracting and bringing into being new actionable worlds, military targeting is about more than the mere act of seeing, taking aim, firing, and destroying a target.

The war on terror would be no different. However, how the targeting assemblage was configured and how it made up the enemy and its targets was distinct. The novel targeting assemblage would produce not only a new type of enemy but also a new kind of war—a war without end, without boundaries and political limits—encompassing the world through newly invasive techniques of knowledge production and destruction.

At present, as this book will show, a number of intertwined trajectories in the histories of military thinking and practice and the sciences and technologies of computing and data analytics are coalescing around military targeting, making it the operational basis for a violent reordering of the world, weaving war deeper and deeper into the fabrics of everyday life and our planetary lifeworlds. From space to seabed, from the tiniest building blocks of the human body to the large-scale connective tissues and infrastructures of social and political life, the rhythms of urban life to the flow of data, the contemporary world of targeting is a world in which everything and everyone, everywhere and all the time, are not only potential targets for action[7], but have already been targeted and are constantly retargeted as new data and reconfigurations engender new relations that make the world measurable, indexed, intelligible, and actionable. We might not always be at war, but we are always targeted.

I see this ongoing transformation as one of the most urgent issues facing us today, with implications that extend far beyond the immediate destruction on the ground. This book is thus an invitation to see beyond damage narratives and extend our analyses of violence into additional spaces and temporalities—into questions about knowledge, power, and dominance, the politics and practices of so-called artificial intelligences and autonomous systems and where we go next. What is at stake here is not only that war and warfare are central to the constitution of social and political orders or that armed conflict and violence are still engrained modes of contemporary politics. More importantly, through the seductive and formative forces of military targeting, the world is designed for war in ways in which the only rational and logical way of thinking about politics and security is through military supremacy, violence, continuous operations, and global domination. In this world, war becomes the definition of normality and the condition of possibility for the future. This is a world made by military targeting for military targeting.

There is thus a pressing need to understand the intricacies and sociotechnologies of targeting and its world-making practices. Such an endeavour starts by recognizing and drilling deep into the operational complexities of military targeting to provide a more nuanced understanding of how violence and effects are inscribed into and standardized through knowledge practices and infrastructures. This is by no means meant to dismiss the importance of reporting on the unfathomable number of victims of military targeting or bearing witness the horrors of war, which, as this is written, continues to haunt humanity. Quite the opposite: this book aims to complement and extend the multiple claims to what matters in thinking about war, intending to ask open-ended questions about violence, not claiming to know what war is or where harm is located. Ultimately, however, what is at stake here is the various ways in which violence operates in our world and its devastating effects on people and societies.

Expounding on the complexities of military targeting, while also accounting for its histories of power, domination, and violence and its sociopolitical situatedness and material specificities, this book offers a broad and potent set of inflection points that aims to open novel avenues for critique through which we can envision alternative and less violent futures.

A World of Knowns and Unknowns

Emerging from the offices of the US Army's Air Corps Tactical School (ACTS) in the interbellum years, military targeting engendered a distinct imaginary, epistemology, and methodology for producing the enemy as a system. Armed with its own particular "lessons learned" from WWI, the ACTS designed a theory of how modern industrial states functioned and sought a way to make these malfunction. The result was a novel vision altering the ways in which the world was understood, conceptualized, and imagined, coupling knowledge production and destruction in ways that turned the problem of war and the enemy into a question of data and data processing.

German and Japan during WWII, the Soviet Union, Iraq, and countless other modern industrial states have all been put under these knowledge machineries, translated, abstracted, and made into an enemy model of causally linked critical infrastructural substrates, distilled down to a set of targets, or "centers of gravity" in military speak, spatially sited and primed for strategic bombing. Believing that the world and the enemy function according to linear cause-and-effect logics, as rule-governed "closed world"[8] systems, the result of these knowledge practices was a model of the enemy that the martial apparatus believed could be used to predict and control outcomes through calculations and computer simulations.

Within this inherently techno-rational view, military targeting's world-making processes literally refigured the world into a system of infrastructurally archived entries rendering the enemy into a static artefact. This was an enemy that belonged to the realm of Hayden's "easy to find, but hard to kill." Not only did this create the illusion that victory could be assured in a scientific manner through the total destruction of key nodes in the enemy system, but it transformed war into a scientific object where the human costs involved were abstracted and the conduct of war was rendered as a purely technical exercise.

Although conventional military forces and their infrastructures try their best not to be detected by the violent sensorium of the adversary, they nevertheless have specific signatures that emit certain signals. They belong in the military category of "known knowns"—a known signature or object in a known location that can be systematized through a functionalist epistemology or what historian of science Peter Galison has described as "destructive functionalism."[9] The former director of the NGA, Letitia Long, described this challenge using the well-known needles in a haystack analogy. During the latter half of the twentieth century, she argues, "the 'haystack' (the Soviet Union or China) was known and with careful attention, we would find the needles. The needles, of course, were objects—tanks, planes, airfields, military bases, missile launch facilities, and the like. Once we found these needles, we tracked and monitored their activity and reported our observations to our customers—policy makers, warfighters, and fellow analysts."[10]

In other words, during the Cold War, both the haystack (location) and the needles (signatures) were neatly delineated; one knew what to look for and where, and one knew how the different elements of the system were related. However, unlike the Soviet Union, or any other modern industrial state with its industrial infrastructures and conventional force structure, the perceived enemy of the "war on terror" posed new challenges for the martial apparatus.

Operating under the assumption that the presumably networked enemy was out there—somewhere, in-the-making, just waiting to emerge from its potential—the US martial apparatus began to see itself operating in an environment in which neither the haystack nor the needles were knowns. This "unknown unknown" became an obsession and worked as a catalyst for several transformations in the US military's doctrinal, technological, and organizational setups as well as war-fighting methods and techniques.

Years before Long's remarks, Donald Rumsfeld had already, and unwittingly, come to frame the epistemological problem confronting the US military. Rumsfeld's oft-ridiculed news briefing on the lack of evidence of a direct link between Baghdad and the supply of weapons of mass destruction (WMDs) to terrorist organizations, spoke about "known knowns," "known unknowns," and infamously, the "unknown unknowns."[11]

Later adding a logical fourth analytical problem (unknown knowns) to the three outlined by Rumsfeld, the US targeting and intelligence community developed a two-by-two matrix depicting the challenge of targeting with four different analytical functions in four different analysis domains (figures 1.1a[12] and 1.1b[13]): the analytical functions of monitoring (in the domain of the known [location] known [signature]), researching (in the domain of the known [location] unknown [signature]), searching (in the domain of the unknown [location] known [signature]), and discovery (in the domain of the unknown unknown).[14]

While this unknown unknown would radically transform the way in which the US military went about war, this was not simply a question of turning unknowns into knowns or "finding" the enemy. The enemy was not merely hiding, camouflaging, or concealing themselves, waiting to be tracked down, found, or unmasked. It was an enemy that could not easily be classified, categorized, and modeled into neatly delineated whole-part structures according to a destructive functionalistic logic. The challenge

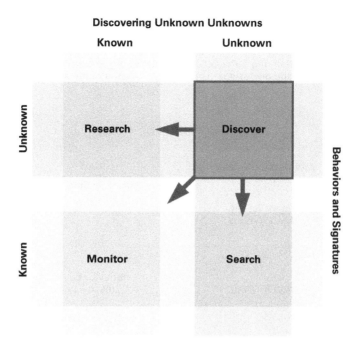

Figure 1.1a

A matrix of the known and the unknowns. *Source*: Letitia A. Long, "Activity Based Intelligence: Understanding the Unknown," *Intelligencer: Journal of U.S. Intelligence Studies* 20, no. 2 (2013): 9.

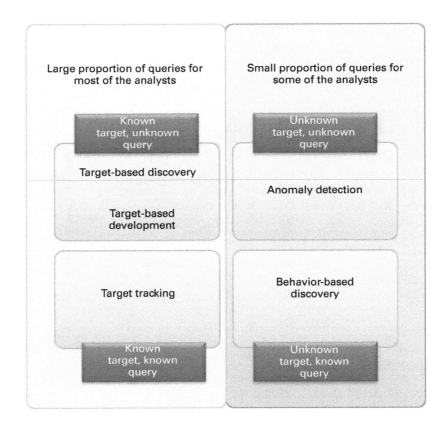

Figure 1.1b

A matrix of the known and the unknowns. *Source*: Government Communications Head Quarters (GCHQ), GCHQ Analytic Cloud Challenges (2012).

was to understand who or what the enemy was, its structures, its parts, and the ties that bind.

Armed with ubiquitous sensing technologies, advanced computational techniques, and machine learning methods, often subsumed under the banner of so-called big data or artificial intelligence, and new targeting methodologies, the US martial apparatus would radically reconfigure the way they produced and operationalized the enemy.

A Diagnostic Screening

While politicians and military professional often portray military targeting as accurate, precise, and a more humane form of warfare, its histories are paved with great paths of destruction. For over a century, targeting's complex world-making forces have made cities and their infrastructures into battlefields and turned humans into objects of carnage. Out of ruined and destroyed urban landscapes across the planet, the visceral and devastating effects of targeting are plain to see—mutilated bodies, shattered societies, and broken communities, living in a state of violence, slow but at the same time potentially abrupt.

Operating at different scales and with different intensities, invasive but not always in the form of violent death and destruction, military targeting is, however, not simply about conducting warfare, seeing, taking aim, and firing, or as the US military likes to sinisterly boast, "putting warheads on foreheads."[15] While military targeting is a force that kills, and a force that has an ever-present potential to destroy, today this mode of world-making operations takes on a particular form that I explore here.

Based on the mass production and processing of data from the ever-expanding digital ecosystem, various techniques of capture, abstraction, and modeling now iteratively correlate and represent the entire world as fluid sets of relationships in which the enemy and the operational environment are seen as spatiotemporal continuums. Modeled and visualized within what is often, in the GEOINT community, referred to as a living multidimensional digital twin of the world, or simply the Map of the World,[16] this is a datascape through which not only knowns can be monitored but through which it is believed that unknowns will be discovered, new knowledge will develop and the "real" can emerge.

Fueled by the invasive techniques and technologies of advanced data analytics and algorithmic models, this is a world in which elements and relations are constantly datafied, mobilized, and operationalized to bring about new realities. In this Map of the World different historical configurations of targeting, martial epistemologies, and analytical problems (knowns and unknowns) combine and fuse to produce the present world of targeting. It is an imaginary, interwoven with a long-standing martial dream of automating knowledge production and decision-making, entangled with complex discourses, justifications, normative considerations, ethical standards, and practices tightly linked to the developments in information infrastructures and technical autonomous systems that offers radical new ways of knowing and operating in the world. The result is a self-sustaining violent "closed-world" view that not only precludes other ways of thinking and acting but that also, through its operational processes, excludes and implodes alternative possibilities and ways of becoming. This is a world forged in the violent histories, epistemologies, operative logics, and practices of military targeting.

To develop its diagnosis of the present, this book takes as its object of study the historical formation of military targeting and its world-making capacities. Highlighting how targeting serves as the core premise that connects successive transformations in martial epistemology and specific forms of warfare, this story connects the interbellum period, through WWII, to the Cold War, the so-called global war on terror, and imagined future conflicts. While military targeting is at the center of this analysis, the broader aim of the book is to offer a historically deep, politically engaged, and empirically nuanced text that makes sense of the magnitude of armed conflict and violence in our times to offer alternative future imaginaries.

Offering this diagnosis on the present, the book engages directly with the gravity of our times, asking how we ended up in a world in which the continual targeting of the world serves as the fundamental form of knowledge production and action. But how did we end up here? What does the world look like from the vantage point of military targeting? How and what kinds of worlds are produced and what futures are imagined, and foreclosed, through military targeting's ubiquitous reproductions and operations of the world? What does military targeting aim at and with what consequences? These are the points of departure that the following adventure sets out from.

An Epistemological History of the Martial Present

Starting off from this diagnosis of the present, I am not only asking in which ways is our present conditioned by the past, but at the same time outlining a research design in which history is used as a means of critical engagement with the present. It is an approach inspired by Michel Foucault's notion of writing "a history of the present"[17]—setting out from a problem of today, beginning the genealogical analysis from a question posed in the present.[18]

How we got here is thus dependent on what "here" is.[19] The opening pages provisionally serve this purpose, offering a diagnostic screening that functions as a guide for "mapping the present."[20]

Military targeting stands at the intersection of multiple genealogies, practices, and concepts, and many stories could have been told. Some have already been told about targeting, and here is yet another one. While the goal is not to conduct a comprehensive genealogical analysis of military targeting, this book is set up as an adventure into the "historical interrogation of the conditions of possibility of things being as they are."[21] A historical analysis carried out with a diagnostic interest does not only have the aim to unearth the past but also, crucially, aims to reevaluate the emergence of the present and, in particular, the many assumptions and ways of operating that lie at the heart of its condition.

It is an approach that focuses analytical attention on the contingent processes of how the past continues to shape the present, being attentive to the details, discontinuities, and recurrences in the emergence of a state of becoming. A history of the present is thus contrasted to traditional historical examinations seeking out origins, structural orders, and discontinuities, or the identification of single ruptures: the why and what of grand historical events.

Repositioning the past in terms of a present diagnostic allows for creating possibilities for revaluing military targeting, whose meaning and value have largely become solidified within dominant narratives, be they critical or orthodox in nature. To situate targeting in a body of relevant historical trajectories is not to suggest that it comprises the logical outcome of these developments. On the contrary, the goal is precisely to suggest that these are not inevitable but, rather, are a radicalization of various trajectories, fused together in a contingent way. The diagnostic interest here is thus a means to reconceptualize and rewrite history, where the end point is not to reconstruct the past but, rather, to reconstruct the present.

This book's interest in the particular knowledge produced through targeting's world-making practices means that this history is concerned with the fundamental epistemological conditions and epistemic operations that surround the present. Approaching knowledge practices historically can prompt us to rethink present epistemological assumptions, concepts, objects, things, and practices. Using history as a diagnostic tool for understanding the emergence of the ways in which the martial knows the world, war, and its enemies is thus also a means through which we, those of us alarmed with the present and future trajectories of martial logics and operations, can disrupt these knowledges and their effects.

Welcome to the World of Military Targeting

Rather than offering a holistic genealogy of the term *targeting*, what is on the menu here is a story that works with the concept of targeting as its own epistemological force of operation. By this I mean a focus on targeting that is not merely applied to shed light on historical instances, events, or transformations but that aims to make targeting a target itself, by mangling historical and contemporary empirics with theory in speculative ways, in which targeting carves out its own empirical paths rather than being guided by particular methods.[22] The result is an exploration that generates unexpected empirical avenues and creates multilayered stories that emphasize the history of targeting's epistemological present, with a focus on its modes of inquiry into the world that produce and operationalize the enemy and make and destroy worlds.

Before we move on, some clarification of terms may be helpful to outline how martial epistemology[23] takes on a prominent role in this story. The adjective *martial* signals, following Alison Howell[24] and others, an attentiveness to the ways in which discourses, practices, technologies, techniques, knowledges, and understandings do not exist or operate in clear dichotomies like military/civilian, war/peace, and exception/normal but, rather, are entwined in or stem from bellicose relations. In other words, *martial* is used in this book as a shorthand for the many entanglements, processes, and practices that do not necessarily come from the military institution, or its many adjacent security and intelligence organizations, but are nevertheless integral for the knowledge produced for, by, or with the military. As such, *martial* is a trope that reminds us that war and warlike relations are not simply derivates of social and political orders but are also constitutive of these orders and their reproductions.

The use of *epistemology*, in this book, does not follow the classic definition of the term as a theory of knowledge but, rather, focuses on the practices and processes of knowledge production.[25] Thus, by invoking the term *martial epistemology*, I make an argument about how the armed forces and the larger martial apparatus try to make sense of the world with the aim of imposing order and predictability to the problem of war's messiness and the unruly enemy. In short, my usage of the term focuses on the martial processes and operations of generating knowledge, turning the analysis toward the intricate relations between sensing and making sense, knowing and acting.

Throughout the book, I am interested in the changing ways in which knowledge and knowledge claims are produced and constructed, defined and evaluated, promised, problematized, rationalized, legitimized, and instrumentalized through various martial "machineries of knowledge production."[26] Or as historian of science Hans-Jörg Rheinberger put it, "the historical conditions *under* which, and the *means* with which, things are made into objects of knowledge."[27] Emphasizing the historical conditions of how we know, historians of science[28] have long been interested in the transformation, development, and process of generating knowledge about objects, or the institutions, practices, technologies, and methods with which they are materialized and how this is initiated and maintained by various actors, tools, and technologies.

Drawing on these insights, my focus is not the content, the specific knowledge created, or what militaries know but, rather, the dynamic processes and reconfigurations of how they know. More specifically, I am interested in how the martial produces knowledge, the material specificities of this production, how these knowledges are operationalized, and the effects and transformative potentials and actualities these understandings have.

In contrast to the martial apparatus's positivist outlook of a world out there, this story also draws inspiration from the tradition of feminist technoscience stressing that knowledge is always contextual, situated, and partial, constructed through means of abstraction.[29] Where you are situated, be it the dark rooms of the Geocell analysts, the air-conditioned containers of the drone operators, the trenches, jungles, or urban landscapes of the battlefield, the offices of war managers, or the executive offices of presidents, is fundamental for how you experience the world and thus also for how you can know.[30] In this way, what is produced is knowledge about not *the* world, but *a* world—the world in which you are spatiotemporally situated.

Drawing on these two separate but intertwined traditions of thought, historical epistemology, and feminist Science and Technology Studies (STS), I argue that this situatedness is not only spatiotemporal but also material. That is, what is and what we know is irreducible to how and by what means we know it. In other words, ontologies and epistemologies are fundamentally intertwined such that the things that come to matter in a world are inseparable from the concepts, theories, techniques, and instruments engaged to know that world. As Karen Barad has argued: "practices of knowing and being are not isolatable: they are mutually implicated."[31]

Although this points to the great debate between realism and constructionism, Lorraine Daston argues that, while scientific objects are constructed, they are "simultaneously real and historical."[32] In other words, objects are real for scientists, but they are also historically constructed. It is in this sense that different epistemic operations can be seen to activate different materialities, congeal different substances and meanings, and enact distinct worlds. What I take from this onto-epistemological vantage point is a need

Welcome to the World of Military Targeting 13

for attending to the material specificities of how things get done, the instruments, techniques, and practices that bring things together and generate knowledge—what I term epistemic operations.

While military targeting is a way for the martial apparatus to know and provide meaning to the world, the epistemological lens developed here also provides a way for a critical engagement with knowledge production. There are different ways of knowing the world, war, and the enemy. Not only are the epistemological assumptions and premises of military imaginaries, discourse, and practices questionable, but the conditions and means of knowledge production are in themselves messy and chaotic, and far from the neat story of a martial apparatus discovering and compiling information about the world "out there."

Questioning the premises of martial epistemology and its epistemic operations, we can engage with military targeting on its own epistemological ground. We can open up and articulate the limits of their self-referential closed-world rationalities through which they justify, juridify, and legitimize their violent practices, and we can undermine their dreams and truth claims by highlighting how their epistemologies are bound up with the histories and politics of domination, power, violence, and importantly, failure.[33] Knowing war in this sense means demonstrating that infrastructures and practices of knowing are at the same time infrastructures and practices of obfuscation, not knowing, or ignorance. This book is interested not only in exposing power and pointing out failures and fault lines but in showing how epistemological imaginaries, practices, and infrastructures support a way of structuring the world that forecloses alternatives, limits the horizons of possibilities, and constrains our understanding of potential futures.[34]

"What's in a War?"

While this book is about military targeting and martial epistemologies, it is also unescapably a book about war. In opening up the door to Etienne Balibar's question,[35] I am not seeking a solution to what war is but, rather, looking for certain inflection points to focus on the quotidian violent practices and techniques that have both social and material effects and resonance beyond the immediate battlefield. Slightly off the beaten academic track, but not alone, this project follows Antoine Bousquet, Jairus Grove, and Nish Shah in arguing that war is not a thing but a question, a becoming rather than a fixed being.[36] What they seek to develop is a radical empiricism in its martial variant—a "martial empiricism" that negates standard abstractions of war, be it from the orthodox or critical literature, by orienting research toward and encouraging us to examine armed conflict through its chaotic empirical processes.

It is an approach that is attentive to armed conflict as an always present condition, a central processual phenomenon in the world with world-making capacities that mutates and transforms in complex and generative ways. In this broader sense, war is constitutive of the (dis)orders of the world,[37] not only a derivate of political orders, and a force that is prior to and beyond the discrete events of clashes of violence. As such, we can argue, war is best treated as a problematic and as an empirical site for investigations into the mutual imbrications and shifting terrains of warfare, technology, knowledge, and politics.[38] From this perspective, it makes sense to direct attention toward heterogenous formations of humans, things, and processes that make war possible and the mechanisms through which war is imagined, assembled, and executed.[39]

As Dan Öberg argues, engaging with military practices in the broader sense of "operating procedures, combat techniques, tactical formations, doctrinal documents and military concepts, the functional roles of technical systems or weapon systems, issues of logistics, staff routines, topography, intelligence gathering, and leadership, to mention but a few [allows] for a more nuanced understanding of warfare."[40] This, he contends, enables a better understanding of how ontologies of war are shaped and performed by military thinking and practice—the everyday activities, processes, and procedures of militaries both on and off the battlefield.

The military institution and those engaged in war—humans and technologies, operators and machines—are preoccupied with making war intelligible and operationalizable.[41] Through a variety of institutionalized practices, the military and its martial organs are, in various ways, trying to rationalize and standardize war and fix its essence.[42] They are, in other words, preoccupied with designing and institutionalizing frameworks that allow for the configuration and operationalization of a set of micropractices and technologies into larger assemblages ready to execute violence on a scale that we commonly refer to as war.

While the military organization, functioning as the institutional manifestation of a state's understanding and practice of war, is central to how warfare unfolds, this is not a claim that knowing the military is the same as knowing war. And one could even argue that, like war, discerning the military, or military power, as an object of study is a futile task.[43] Nevertheless, the organizations of armed forces and the military institution are crucial actors in and of war. Being the organized way in which states are trying to make sense of and provide meaning to war and turn it into an instrument of politics, the paradox is that the military institution stands as the antithesis to an approach to war as becoming. But on the other hand, these epistemic operations and practices also reveal that war is not a thing but a messy process of becoming.

Engaging in these processes can thus highlight the ways in which imagining, assembling, planning, and conducting war, the primary tasks of militaries, are not only derivatives of social and political orders but also shape and reconfigure these orders and are themselves shaped by war's volatile relational characteristics. What war is, for them, is thus a correlate of what militaries and the larger martial apparatus know about the world at particular historical junctures.

Understanding war as shaped by the standardized and rule-bound practices of military bureaucracies, Astrid Nordin and Dan Öberg argue for turning our attention to studying "war as processing." By focusing on processing, Nordin and Öberg are interested not so much in the standardizations and bureaucratization of war per se but, rather, in the ways in which contemporary war-fighting concepts are designed to perfect war and make it more efficient through the seamless integrations of operational processes. For them, military targeting is one such process that describes how warfare should be conducted, producing a model that makes war into a process ad infinitum. Turning war into reiterative processes, they argue, produces a repetitive battle-rhythm, and a model of war that when operationalized turns war into warfare and warfare into the technical realization of a preplanned script. Consequently, they contend, the process itself becomes the agent, in which the operations are not only the means but also the endpoint of war.[44]

A commitment to follow the empirical complexities of war and studying war as process does not promise an understanding of what war is, but it is a stance that seeks a more nuanced understanding of war, its mechanisms, and its effects while at the same time acknowledging war's chaotic unfoldings. More importantly, however, this commitment underpins a critical project that seeks to challenge how the phenomenon of war is apprehended, how knowledge about it is produced, and what it means to know war.

Although this critique of war emerges from episodes of war, the aim is to rethink the ever-present condition of war, or more precisely, the understandings that sustain war as a persistent feature of our world. Foregrounding the martial quest to tame the radical and chaotic aspects of war, to overcome fog and friction, we can at the same time use war's chaos to challenge these operational logics, to demonstrate the limits of martial epistemologies' quest to standardize and fix the chaotic realities of the ground, showing not only that they are phantasmatic dreams but that there are different ways of knowing and addressing the phenomenon of war and violence.

Know Thine Enemy

What war is, how to conduct warfare, and who the enemy is are questions that long predate the modern conception of military targeting, and there are many different ways in which the martial apparatus seeks to know. What different martial ways of knowing have in common is that they work from a positivist assumption that the world "out there" is ripe for the picking, that knowledge from it can be extracted if the right tools are available. But, as we have seen, neither the world nor the enemy is a natural kind that simply can be found "out there"; it must be, to reiterate, "made up."

Historically, Mathias Delori and Vron Ware argue, several different "faces of enmity"[45] have been fashioned: from the racialized, demonized, and politically named enemy Other[46] and the dehumanized anonymous mass as targets for strategic bombing[47] to the cybernetic human-machine centaur enemy[48] and infectious diseases[49] to the contemporary individual Other as signatures and anomalies in vast datasets.[50] What all these have in common is what Edward Said once told us was one of the central features of Orientalism: "a proclivity to divide, subdivide, and redivide its subject matter."[51]

There is a long historical trajectory in the critical scholarship on enmity and representations of the enemy aiming to show how the process of publicly naming an adversary or threat is a prerequisite for the permissibility of action. This body of work has studied the social construction of enmity and processes of "othering" through identity politics[52] and the construction of threats, dangers, and risks to various referent objects through different securitization processes[53] that not only promise the destruction of enemies but also biopolitical governance[54] and killing to make life live.[55] While there is no shortage of critical literature on the formation and framing of enemy figures and differentiating "us" from "them,"[56] this scholarship is largely rooted in discursive theories focusing predominantly on political elites' naming of Others.

However, discursively made enemies do not come complete with a set of operational procedures and actionable objects through which the military can simply execute its orders and conduct warfare. Armed forces and security professionals also make up their own figure of the enemy, and a number of recent studies have shown how different subjects and objects emerge from various martial practices.[57] While existing in separate fields of discourse and practice, these figures do not stand alone or completely separated from the discursive Other. Rather, they are entangled, coexisting and constantly reinforcing and influencing each other.

How the enemy is made up is important for how militaries (re)imagine war, plan for armed conflict, allocate investments, produce weapons and other materiel, assemble their forces, (re)construct operational environments, and wage specific forms of warfare. However, for warfare to take place, a specific actionable understanding of what the world and the enemy are needs to be produced—drawing boundaries between friend/enemy, us/them, self/other, and inside/outside by classifying and categorizing, identifying, and sorting[58] the good from the bad, the "'irreconcilables' from the 'reconcilables'"[59] and the risky and threatening from the safe and secure. Although never as clear-cut as these dichotomies portray, or simply a matter of drawing distinctions, such an "architecture of enmity"[60] is nevertheless a condition for the possibility of warfare. These are multifaceted processes that entangle discourses, practices, and technologies in complex but specific ways.

Returning to Ian Hacking's notion of "making up people," it makes sense to also orient our analysis toward the other ways in which the enemy is produced—the practices, techniques, and processes that make the enemy into operationalizable models and objects. So, what does the martial apparatus see when it looks for the enemy?

"Enemies were not all alike," historian of science Peter Galison argues in his investigation of how different enemy figures emerged through scientific advances and their intersection with military thought and practice during WWII:

> In the killing frenzy of World War II, one version of the Enemy Other (not Wiener's) was barely human. . . . These monstrous, racialized images of hate certainly presented one version of the World War II enemy, but it was by no means the only one. Another and distinct Allied vision held the enemy to be not the racialized version of a dreaded opponent but rather the more anonymous target of air raids. This enemy's humanity was compromised not by being subhuman, vicious, abnormal, or primitive but by occupying physical and moral distance. . . . But there is yet another picture of the enemy that emerged during World War II. . . . More active than the targeted, invisible inhabitants of a distant city and more rational than the hoardelike [sic] race enemy, this third version emerged as a cold-blooded, machinelike opponent. This was the enemy, not of bayonet struggles in the trenches, nor of architectural targets fixed through the prisms of a Norden gunsight. Rather, it was a mechanized Enemy Other.[61]

While Galison concentrates on the famous scientist Norbert Wiener and the making of the cybernetic other, his astute observation highlights that the figure of the enemy exists in a variety of forms simultaneously, manifested in multiple ways, ranging from the discursive or narrative, through figurative and metaphorical, to mathematical and computational representations and material realizations.[62] What interests both Hacking and Galison, and what is important in this book, is the ways in which ideas, concepts, things, objects, people, and so on "come into being" in particular contexts and how these are open to historical change.[63]

Figures of the enemy appear in different fields of practice, linked in various ways to the justifications, legitimizations, and juridifications of violent practices, and they activate different forms of action and violence. Enemies exist outside of the field of military targeting, but they also exist here, in important and underrepresented ways. This is not to say that the justification and juridification of war and violence are not important, as they certainly are,[64] but this book offers a complementary view of making up the enemy that sits between the discursive production of the enemy and the death and destruction on the ground. What I am trying to map in this book is how knowledge about the enemy is produced and operationalized. At stake here is thus not just how justifications or juridifications of war are produced but how martial epistemologies and meaning-making are inscribed with war and violence.

Although it would be impossible to account for the past hundred years of Western warfare without some mentioning of data production and computing, this is not a question of "what difference did computers make?"[65] or any other technoscientific tool or technique, any more than it is just a question of how the enemy is narratively or discursively produced. Thus, rather than asking how humans classify things, or how instruments make objects, it seems more appropriate to ask how human-machine configurations classify us and the things around us, and what difference specific configurations make at distinct historical junctures: how did historically contingent and emerging configurations perform certain epistemic operations to abstract, model, and operationalize worlds and the enemy? And to what effects?

As the following pages will show, the enemy emerges not as a fixed ontology but as an effect of historically distinct configurations and their ongoing processes and practices that systematically and structurally "identify and sort,"[66] classify and categorize, localize, model, and make up that of interest. This means that the emergence of distinct historical understandings of enmity and the contemporary world of military targeting are contingently dependent on the configuration of various elements that do not have a single origin but are multiple processes of historical emergence entangling civilian, commercial, and military spheres.

Making Up the Book

After this introductory chapter, the second chapter introduces a set of concepts, methodologies, and investigative lenses that are used to guide this book through its historical interrogation of military targeting and its world-making capacities.

The chapter starts off with a critical reading of the US joint targeting doctrines, showing how their relational logics assume a world built by whole-part relationships and how this prescribes particular ways of operating in this world. Reconceptualizing military targeting as a knowledge practice, the chapter goes on to argue, turns the analytical gaze and investigative lens from perception to cognition and the many different epistemic operations that process information to make up the enemy. In order to analyze these complex knowledge practices, I develop a particular end-to-end approach that is designed to traverse the spatiotemporal scales between politics and military imaginaries, particular configurations of humans and machines, the concrete epistemic operations these conduct, and the violent effects on the ground. However, this is, above all, a methodology that seeks to immerse itself in the empirical complexities and details of military targeting at four different historical junctures.

Spanning a timescale from the interbellum years, through World War II, the Cold War (chapter 3), the Vietnam War (chapter 4), the war on terror (chapters 5 and 6), and into the future, with a focus on the US military and its wider martial apparatus, each of the four empirical chapters explores how military targeting was configured, how it produced and operationalized the enemy, and how this reconstructed the operational environment and engendered specific transformations in warfare and to what destructive and violent effects.

The chapters are loosely structured around the two-by-two matrix from the introduction (figure 1.1a and 1.1b), each of them exploring the challenges of the knowns and unknowns; the analytical problem of known knowns (chapter 3), known unknowns and unknown knowns (chapters 4 and 5), and unknown unknowns (chapter 6) and the analytical functions of monitoring, researching, searching, and discovering. Being loosely structured around these functional practices provides the empirical chapters with an artificial matrix through which to analyze military targeting at different historical junctures—engaging directly with what the martial apparatus defines as a problem and how this apparatus attempted to solve it.

I say "artificial matrix" because there are substantive overlaps and continuities in the practices and processes identified and analyzed, and because the "known" category is only known after it has been defined, stabilized, named, and put into functional relationships. The matrix is thus not meant to present a succession of historical stages, although they are largely chronological, but rather, serve as an analytical lens to distinguish among analytical problems, the epistemologies they rest on and engender, and the ways in which configurations are designed to solve these problematizations.

While it is possible to trace different military epistemologies to specific historical moments, I do not suggest an epochal analysis of radical transformation. Rather, I propose to see how these different forms combine and emerge together in complex relations to produce the present. Moreover, the choice of these historical periods is largely based on times of transition, highlighting the various ways in which continuities, accelerations, experimentations, and disruptions shaped, changed, or altered the way targeting was configured and produced the enemy. Targeting is thus seen as transhistorical, moving through historical moments, connecting successive transformations in martial epistemology, the structuring of enmity, and specific forms of warfare, with its own particular violent effects.

The aim is not only to isolate particular historical martial epistemologies that give rise to the field on intelligible things, but also to map the ways in which that field fails, the instants of its disruptions. Wars, of course, tend to bring about a questioning of knowledges and understandings, being a highly disruptive, messy and generative phenomenon. However, the sites where a particular epistemology "fails to constitute the intelligibility for which it stands"[67] and a new one arises do not mean the failure of a martial epistemology to produce knowledge but, rather, a failure to produce meaning in a new ecology.

Reframing military targeting as a martial epistemology and mapping its emergence from systems theory and foregrounding its knowledge production, the book's experimental narrative emphasizes different histories, problematizations, discourses, instruments, and techniques than do other studies of military targeting and genealogical accounts of contemporary warfare. This positions this enquiry not only as part of the profound relationship between science, technology, and warfare but, more specifically, as a relationship between military imaginaries, martial epistemologies, knowledge-production practices, computational instruments, and war. It thus focuses not only on the scientific ideas that have been recruited to think about war and the organization of warfare,[68] the way in which war's scientific and technopolitical understandings and representations make it manageable according to managerial strategies,[69] or the cybernetic origins of Cold War military discourses, technologies, and practices[70] but on the way in which specific targeting configurations and their data logics and techniques work to produce and operationalize the world.

Bringing attention to how imagining the world, producing the world, and operationalizing the world are always wrapped up in each other and profoundly embedded in sociotechnical systems and processes, the chapters inscribe the contemporary world of targeting in a long-standing and concerted martial effort to know and operationalize the world and the enemy. Drawing detailed attention to military doctrine and methodologies, statistical thought and practice, the mathematical and computational techniques of data production, processing and modeling, and various technologies from human to technical sensors, archives, and punch card devices to the vast databases and machine-learning algorithms and so-called AI of today, the chapters provide a detailed mapping of a hundred years of transformations in epistemic practices and techniques of translation in relation to machines.

This story about military targeting begins at an inflection point in history when a number of military discourses, practices, and technologies came together to form the practice of military targeting. I locate its emergence in the 1930s alongside the invention and use of the airplane and the entanglements with what Hunter Heyck has termed the "age of system"—a specific way of seeing the world and modern societies as a system composed of smaller subsystems.[71] Mapping the various reconfigurations of targeting through the interwar years, WWII, and the Cold War, chapter 4 explores how the enemy was produced, processed, analyzed, and archived, making the known knowns part of the vast archive of what was known as the Bombing Encyclopedia of the World. This enemy-as-a-system epistemology saw to it that the enemy is no longer simply a collection of armed forces but, rather, a collection of infrastructural nodes vital for the functioning of modern societies, turning war into a rational calculation and cities into battlefields.

While nuclear war planners were busy trying to figure out how to best wage a suicidal war against another nuclear power, the Vietnam War would, as chapter 5 narrates, represent a novel problem for the US martial apparatus. Up against a "faceless enemy" who did not offer any of the infrastructural targets the US martial machinery had been configured to produce, they designed and engineered various experimental reconfigurations to capture the unknown knowns and known unknowns. Armed with the novel computer and various statistical and mathematical techniques to make the environment computable and translatable, the Hamlet Evaluation System (HES), the Igloo White, and the Phoenix programs produced and modeled the enemy at various levels, from the rural population, social relations, and behavior of individuals to sensory signals in the jungle. These were configurations that supported the vision of war as winnable through managerial and technoscientific logics and rationalities, turning Vietnam into an experimental laboratory, bringing terrible devastation to the peoples, the sociocultural environment, and politics in Indochina.

Drawing lines from the war in Vietnam, chapter 6 takes us back to the introduction with a particular focus on the F3EAD (find-fix-finish-exploit-assess-disseminate) targeting methodology, and the novel computational practices, instruments, and techniques marshalled and inserted into the targeting configurations to produce and uncover the unknown knowns and known unknowns. Showing the peculiarities and complexities of this methodology and its epistemic operations, it will be argued that this transformed targeting from planning based on knowledge to an emergent understanding of the enemy based on learning through recurrent violent feedback loops, creating its own self-sustaining logics of perpetuation and expansion, where military targeting becomes an end in itself.

Chapter 7 comes back to the introductory diagnosis showing how we are now living in a world in which everything and everyone, everywhere and all-the-time, has already been targeted and is constantly retargeted. Examining the algorithmic techniques of activity-based intelligence (ABI) and its focus on anomalies, the chapter will highlight how the martial apparatus is now merging the techniques and enemy models from the previous chapters with unknown unknowns. In this contemporary targeting regime of experimental tinkering and tailoring, military design thinking and the computational ethos of trial and error converge, redefining instrumentality away from goal-oriented planning or control and toward open-ended violent trial and error, harnessing emergent potentials in which there are no ends, only future beginnings.

The concluding chapter revisits the diagnosis of the present to critically reflect on the kinds of worlds contemporary targeting is continuously (re)configuring, (re)constructing, and (re)ordering. A particular concern is raised about the move toward the ethos of trial and error that is radically redefining instrumentality and turning the future into a world of constant military operations. The chapter highlights some of the acute ethico-political issues that cut across the contemporary (re)configurations of military targeting and its technoscientific logics and practices and shows ways to reimagine a different world.

2
Military Targeting as Epistemology

While military targeting has received its fair share of attention in recent years through drone strikes and so-called targeted killings, this book treats targeting in different ways than is common in media, professional, and scholarly circles. Although these prevalent representations of targeting—often focusing on its violent effects on the ground or questioning official claims related to the legality, accuracy, and precision of these operations—have made important interventions into the proclaimed legitimacy of contemporary warfare by exposing the failures of technological fixes, there is a relative absence of detailed investigation into military targeting.

Targeting has for the most part been hidden in investigations of intelligence, air power, strategic bombing,[1] and drones;[2] in legal and ethical debates on what constitutes a legal target;[3] and in more traditional and technical military discussions such as how to hit a target and what effects this has on the objective of the war or on the effectiveness of targeted killings as a policy.[4] Thus, while many scholars have recently shed light on the importance of the processes of targeting for contemporary warfare, military targeting, as an object of inquiry in itself, let alone its history, has rarely been the focus of attention.[5] This is puzzling given the centrality of the methodologies, infrastructures, processes, and practices of targeting to how wars are imagined, planned, organized, and conducted.[6]

Historically, targeting was primarily a tactical decision-making exercise aimed at effectively destroying enemy forces and materiel to produce effects on the immediate battlefield. This changed with the introduction of the airplane. Theorists and practitioners saw this new technology as a way to achieve effects beyond the battlefield and at all, so-called, levels of war: strategic, operational, and tactical. However, this was not only a question of the airplane but, rather, a moment in time when a number of interlinked developments in systems theory, datafication of the world, data management, and computation coalesced to engender the epistemologies of military targeting. Over time, targeting developed into a knowledge-producing approach, increasingly intertwining meaning-making and decision-making.

Today, targeting functions as the operationalization of doctrinal and administrative military processes; it allows specific methodologies and methods of operating, focused on constructing targets, selecting and prioritizing targets, and tasking military action to achieve military goals.[7] Operating as a bridge between the ends and means of warfare, producing and selecting targets, targeting is essentially about military operational thinking and conduct.[8] As such, targeting manifests itself as the Clausewitzian politics of war:[9] instrumentalizing violence, prescribing how forces should be structured and employed to achieve ends, guiding warfare, and thereby assigning meaning and logic to war. This positions military targeting as an imagined rational, technoscientific, and epistemological mode of inquiry, a way of operating, and a functional worldview in which actions have entirely predictable effects and desired outcomes are not only achievable, but also, it is assumed, predicable.

In military parlance, targeting is described as the "application of force to produce an effect on selected battlespace elements,"[10] but in order to apply "force" to produce an "effect" on "elements" within a "battlespace," one necessarily has to know what these elements are and where they are located. In other words, both the elements and the battlespace must be produced and constructed. The processes and practices of targeting thus provide a way of translating the world into a set of actionable elements within a designated space.

As we can see from figure 2.1,[11] the US joint targeting cycle is usually portrayed as six sequential steps, from the objectives of the commander all the way to an assessment of the effect produced by the use of force. Given the focus on martial epistemology and epistemic operations, the ensuing empirical analysis of targeting will predominantly focus on what the joint targeting cycle describes as "phase 2, target development and prioritization," or what this book calls the production and operationalization of the enemy. Figure 2.2[12] breaks phase 2 down further, providing a targeting taxonomy that orders the enemy, its capabilities, and its elements hierarchically according to a systems perspective—the set of interrelated components that perform a specific function or capability of the enemy.

Crucially, the Joint Targeting Doctrine states, "While a single target may be significant because of its own characteristics, the target's real importance lies in *its relationship to other targets* within an operational system."[13]

What figure 2.2 and the above quote emphasize is that military targeting is fundamentally a distinct relational outlook on the world. Here, military targeting assumes that the composition and patterns of the world are underwritten by whole-part and part-part relations that can be grasped as systems and made knowable through abstraction.

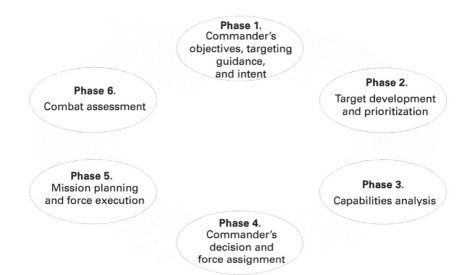

Figure 2.1
US joint targeting cycle. *Source*: US Joint Staff (2018), *Short Joint Publication 3-60, Joint Targeting*, II-4.

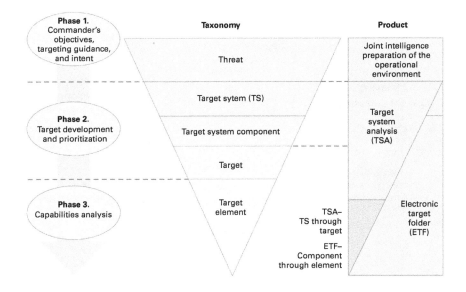

Figure 2.2
Target development relationship. *Source*: US Joint Staff, *Short Joint Publication 3-60, Joint Targeting*, II-6.

"Abstraction," cyberneticians Arturo Rosenblueth and Norbert Wiener tell us, "consists in replacing the part of the universe under consideration by a model of similar but simpler structure."[14] More than an apparatus of distinction[15] differing between categories of combatant/noncombatant, friend/enemy, and military/civilian, targeting is a process of gathering otherwise disparate elements into an articulated whole, distilling a series of relations into a "real" entity. Assuming the world "out there" exists independently of the means by which the world is abstracted, this is a positivist outlook in which reality and the world are thought of as accessible and readily available to be discovered through the right sociotechnical techniques of questioning and capture. It is an approach grounded in a kind of realism that assumes that the real emerges from data and that knowledge emerges from the systematization of observations.

Through multiple operations of data production, data processing, and interpretation, military targeting seamlessly moves and interprets data from generalities (enemy) to particulars (elements) and back again. Being about parts, wholes, and relational ties, targeting defines relationality by homogenizing objects down to their functional relationship to other objects in the system. It is a process reminiscent of Gadamer's "hermeneutic" circle: "The anticipation of meaning in which the whole is envisaged becomes actual understanding when the parts that are determined by the whole themselves also determine this whole. . . . The movement of understanding is constantly from the whole to the part and back to the whole."[16]

The result of these operational processes is the emergence of the enemy as an abstracted model—a model descriptive of the world, but at the same time prescriptive about how to operationalize this world. As a simplified representation of a "real-world" system, the enemy appears as a coherent entity, which can only be rendered visible and acted upon through numerical and calculative processes and techniques. Thus, targeting does not just compile entities associated with "enmity," but it creates and forms the enemy as a coherent entity in nonnarrative computational form.

While appearing as a coherent object, the existence of this modeled enemy is dependent on a complex environment that includes historically situated assumptions, values, and biases; the data produced and made available; the computational infrastructure to process the data; different methods for analysis and modeling; and experts and tools for interpreting and visualizing the model output. While models do many things, the most important aspect here is that they make the enemy and the world visible in a systematic and comprehensible way—a "similar but simpler structure."[17]

But like any other model or "simplification,"[18] these are not only representations of complex systems, but active components in how worlds are seen, problems understood and solved, experiments conducted, and action generated. In other words, models are, as Rosenblueth and Wiener argue in the context of the natural sciences, both abstract representations and enablers for scientific inquiry.[19] Military targeting produces particular models of the enemy that are not stable structures but ontological distinctions specifically designed to function as tools to be manipulated for further inquiries, or mobilized for warfare.

In this sense, models work as interfaces to the sociopolitical world, operative devices to investigate and intervene in the world, and operative agents that modify the world.[20] Targeting is thus operational rather than representational. Here, targeting becomes what Massumi has theorized as "operative logics,"[21] a "force of attraction"[22] around which data and operations, abstractions, and models are being made and maintained, functioning as ontogenetic forces that bring about realities and an actionable model of the world.

Conceptualizing military targeting as a martial epistemology, the empirical investigation of this book is homing in on the epistemic modeling operations that are historical and political, technological and infrastructural, always embedded in practice, which is at the same time both organizational and informational and situated and contextual. Military targeting is performative, but not only in the linguistic and speech act sense. It is, to paraphrase Donald MacKenzie, not only a camera reproducing a world picture, but an engine designed to operate in and on the world, making it, molding it, transforming it, and destroying it.[23]

This martial epistemology is shaped by historical, cultural, and political factors, but at the same time its epistemic operations are contextual and situated in the sense that how the enemy is produced, represented, and operationalized varies according to the situation and fluctuates according to the relational conditions linked to goals. For these reasons, the knowledge, choices, and decisions produced through military targeting cannot only be analyzed as the establishment of facts, boundaries, or categories according to some predefined strategic rationality or overall military imaginary but must also be viewed through the actions they generate and their operational goals and effects. Although closely related to traditional military intelligence, targeting does not solely center around problem-solving or what Robert Merton once called "specific ignorance,"[24] lack of knowledge that you can pin down (known knowns), that aims to uncover hidden truths or just the hidden, but is experimental and exploratory, with the goal to produce an understanding of the enemy for tactical and operational effectiveness.

In addition, situated between military intelligence and military operations, targeting is an iterative learning process, coupling knowledge and action, in which both feed off each other. Through modeling and simulations, destruction or operations, new information is gained, meaning that learning and then knowledge are thought of as appearing through the processes of formalized operation.

By transforming how the world is sensed and apprehended, abstracted, and known, targeting creates simplified structures that not only reduce the complexities of the world but create models in which simplified linear causal paths can be computed and predicted. These pathways are ridden not only with death, destruction, and despair but also with failures to achieve political goals. That calculated output in the form of destruction is not the same as political outcome seems to be a lesson that is never learned, and the failure to understand this repeatedly put decision-makers on the road to warfare.

From Perception to Cognition

This story of military targeting has so far indicated that its epistemic operations are more than perceptual practices. What I have alluded to is the importance of the process of interpreting computational information, or data, and turning data into meaning according to the relational logics of the martial apparatus. Martin Coward, debating the changes in US military thinking and practice in the 1990s, notes that the centrality of data to the martial machinery means that sensing, knowing, and meaning are not just ocularly produced or perceptually formed but are engendered through various filtering, translation, and meaning-making processes, in which data and computation are essential: "Where the logistics of perception evolve from the line of sight and are largely ocular in orientation, data constitutes a mechanism for mediating between that which cannot be perceived and that which can be understood. As such this is less perception than cognition."[25]

While Coward does not elaborate further on this, or what cognition entails, the centrality of data to the war machine is difficult to understate. In fact, it is challenging to understand or at least account for the last century or so of warfare without considering the fundamental importance of the datafication of the world and the increasingly ubiquitous information infrastructures for knowing the world and its attendant configurations of humans and machines.[26]

For the past decades, one of the most fruitful ways of critically interrogating contemporary warfare and military targeting has been to engage with different genealogies of military vision systems,[27] the narrowing of vision and boundary-making of "scopic regimes"[28] and various ways in which the aerial gaze, from aerial reconnaissance to the hunter-killer drone, produces geography in a specific way, engendering a politics of verticality and remote management,[29] with particular effects on the governing of populations.[30]

Although this literature has focused on targets and, to some extent, targeting, it has done so primarily through critically opening up the aerial gaze as a cultural and an imperial/colonial way of seeing alongside its connected practices of aerial bombing and the military "sensor-shooter" problem.[31] Derek Gregory, in his reading of bombing's history, has, however, explicitly argued that the air target is made and brought into being through complex procedures of identification, calculation, and translation,[32] leading aerial warfare to require a new form of technique for producing the battlespace that is inherently calculative and intelligence rich.

Importantly, as highlighted by the literature on the aerial gaze, the production of these targets should not only be seen as a "product of developments within technologies of vision, but also developments within the epistemologies of geographical knowledge production," scopic regimes of visuality, and sensor technologies.[33] Although recognizing how targets are increasingly understood as systemic and relational and to a large extent infrastructural,[34] adding up to a representation of the enemy Other, this literature is in large part interested in the cultural and geographical production of battlespaces and targets, often with an explicit focus on the effects of bombing and not how this constitutes specific understandings and structures of enmity. These practices are, however, not only culturally situated but also fundamentally technically and functionally situated.

It should come as no surprise that that differently situated military professionals and politicians "see" and "know" war differently, but how and why they see and know it differently is important. Gregory, for instance, uses the term *epistemic rupture* to argue that the techniques of aerial reconnaissance and optical cartographical technologies used during WWI produced an abstract space and alternative world of charts and graphs that were not only very different from the soldiers' perception from the trenches, but in large part supplanted the reality on the ground.[35] Anna Danielsson, pushing this argument further, argues that it is not simply a question of different spatial situatedness or a matter of different perspectives, the air or the ground, or the different military branches, but is in large part an onto-epistemological question of different forms of world-making. According to Danielsson, operational environments are brought into being in and through epistemic operations as much as knowledge production is constituted by that very same environment.[36]

What both Gregory and Danielsson show is that epistemic operations are processes that are situated in particular socio-material environments where perception is structured in ways that matter for the knowledge output and interpretation. Battlespaces are not only perceived in different ways but brought into being in different ways.

Oliver Belcher also contrasts the view from above with what he has termed the "view of below." Analyzing the US computerized technologies used by the Hamlet Evaluation System during the Vietnam War, Belcher argues that these were not designed to simplify complexity as in, for instance, James Scott's argument of "seeing like a state,"[37] or the aerial view, but rather, allow for the complexity of population dynamics in Vietnam to be captured, translated, and modeled. That is, the various databases constructed through producing, processing, and analyzing and then archiving data served as a way to both contain and enact complexity while making it work for a particular purpose. This, he contends, enabled an instrumental horizontal perspective that engendered a different form of knowledge and military action.[38]

Drawing on these different ways in which military knowledge is produced, I would like to return to the notion of cognition, as I see it as crucial to understanding the epistemic operations of military targeting. Cognition complements and extends the literature on the aerial gaze, sensorial technologies, and the epistemologies of geographical knowledge production in important ways.

Here, I draw attention to Katherine Hayles's notion of "cognitive assemblages"–those assemblages that perform cognitive operations, understood as processes that interpret information in context and connect it with meaning[39] (in simpler terms, assemblages consisting of sensors that produce data or *input*, the analytical techniques and technologies used to *process and analyze* these data, and the methods and practices used to model and interpret the *output* of these operations and connect them with meaning).

Crucially, for Hayles, cognition is not an attribute but, rather, a recursive process and reciprocal activity that unfolds with an environment. Unlike thinking and perception, cognition requires systematic interaction and navigation with the environment to interpret information and connect it with meaning.[40] It is, therefore, more than a perceptual event and more than thinking, as meaning develops in context and the choices mutate and adapt as the various micropractices of cognitive operations confront the environment. Following Hayles, the key to capturing the martial epistemologies of targeting is a focus on how specific historical configurations of humans and machines turn data into meaning, turning worlds into models and models into "accessible targets."

Decentering the visual way of war, military perception, and the martial gaze[41] and moving military targeting toward questions of epistemology and legibility show that the enemy is not the result of a single sensory event or of how the world is perceived in any given moment, but, rather, is the result of the recursive cognitive operations—the operations that turn perceptual events into data, data into calculations, and calculations into meaning, decision, and action.

By moving the scholarly and popular lens from killing to knowing and perception to cognition, we can bring attention to how imagining the world, producing the world, and operationalizing the world are embedded in sociotechnical systems and processes. As such, military targeting is not simply a methodology, a perceptual practice, or a technique to solve the problem of war and the enemy, but a particular martial epistemology for producing and operationalizing the world and the enemy.

From this perspective, military targeting is as much about analytical power as it is about sensorial powers, about intelligibility and translation rather than surveillance. It is also more than a classification practice, sorting and ordering things into categories,[42] because it inscribes meaning and context to objects, which shapes and determines not only decision-making and acting but also further classifications based on the input/output of the targeting operations. Central to this is the sensor-analysis-output triangle of data production, processing, and interpretation, translating the world into data and data into worlds.

Making Sense of Epistemic Operations

In everyday language, data or information is often thought of as collected. However, sensors not only enhance human sensing but are already structured and trained through the logics of the targeting configurations; they embody values, beliefs, and desires relative to those of their designers and users. In this book, sensors are conceptualized as either humans or technological objects that are trained, designed, engineered, and manufactured to capture, record, and transmit data. Sensors are, to use Sara Ahmed's words, "oriented" in certain ways via sociotechnical arrangements and spaces to sense and attend to some directions, objects, and situations, but not others.[43]

The means of sensing are therefore never neutral. Not only are they products of specific historical, political, and technological contexts and imaginaries but they are also undergirded by particular epistemological assumptions, with their own built-in biases, opacities, and partialities about what constitutes relevant and good data. In this way, data, understood since the mid–twentieth century as computational information, are not simply "collected" or merely "given," but are produced, manufactured, and modulated in particular ways.

In the martial world, sensors can be anything from human interrogators, observers, or spies to satellites, cameras, radars and lidars, acoustic buoys, microphones, wiretaps, or pieces of software that "scrape" the digital ecosystem. These are vital for military targeting because they enable the environment to become computational,[44] "datafying"[45] the context in which they are placed, be that the ocean, urban terrains, jungles, space, the internet, phone traffic, or interrogation rooms. Thus, sensors gain political and epistemic relevance, enabling and constraining what is and is not relevant data to be analyzed, reducing the possible multiple ways in which the world can be translated into "reality." How sensors shape and are being shaped by the spatiotemporal environment they are inserted into and the processes, relations, objects, and events they attempt to make accessible is, however, an empirical question.[46]

Once data are produced, labeled, and categorized, they are processed to generate patterns and relations that eventually can be churned into a model. Such analytical techniques range from the simple list to more complex statistical modelling of linear regression, logistical regression, and Cox models to more novel forms of algorithmic modeling using machine learning techniques such as Random Forest and so-called deep-learning neural nets.[47] The different ways of analyzing data are both historically and contextually important, as they inform the design of sensors and the production of data. Also, the different computational techniques process data in different ways and generate different outputs that condition the interpretation of the outputs. While the choice of tool or technique for data analysis defines the data, it also determines what kinds of output can be generated, conditioning how these are interpreted and connected to meaning.

Another important element in the sensor-analysis-output triangle is the various targeting archives and databases that shape and support these processes. These are, in John Durham Peters' words, "containers of possibility."[48] Databases are not simply spaces of permanent ordering of objects, preserving memory, but wider information networks allowing sharing, reusing, remixing, modification, and importantly, modeling. They are thus best conceptualized as "active site[s] for the execution of [epistemic] operations"[49] in which their built-in functionalities and logics frame the relationship between space, time, matter, and memory through which the enemy model can be recombined and remodeled and made actionable in different contexts.

While databases serve as the condition for modeling the enemy, the models themselves operationalize the databases. The various databases constructed through the processes of producing, processing, analyzing, and archiving data serve to simultaneously contain and enact complexity. This is not to say that targeting does not rest on abstractions but, rather, that the vast databases engendered through targeting are not static repositories of simplifications.[50] While databases and techniques and technologies for data capture in large part determine what kinds of data can be produced and stored, the ability to query and endlessly recombine the data means that what can be analyzed and discovered of the world and made actionable depends on the contextual processing and interpretation of this data. The processes and epistemic operations of targeting can also produce different models of the enemy from the same data. In this sense, the weaving together of data is a scientific or technical endeavor but also, and importantly, a cultural and political one.

Such a view of the data-processing triangle can help make visible that violence is also located in important ways in the production and analysis of data and in the complex epistemic infrastructures and techniques of translation that make the world computable, modelable, and actionable. In other words, and as the empirical chapters will show, how data are produced, how they are processed, and how these are interpreted and connected to meaning matter for how things are mattered or put into relations and made actionable.

Configuration, Imaginary, and Agency

While there are no blueprints for how to conduct "histories of the present," the conditions of possibility for phenomena and processes are tightly linked to the question of how different heterogenous elements, be they discourses, practices, technologies, or material forces, come together and work at specific historical junctures. Lucy Suchman's concept of configuration[51] is especially important here, as it draws analytical attention to the way in which specific cultural imaginaries[52] are built into sociotechnical systems via design and engineering practices, asking how the various components of an assemblage are figured and what that means for the ways in which they are put in relation, or figured together. Suchman thus offers a way to study how things come together, not as pure contingency or response to an urgent need, as per Michel Foucault[53] but with a focus on how specific imaginaries are materialized and made actionable through the design and engineering work needed to make various elements compatible.[54]

It is useful here to think about the concept of military imaginaries used to describe the ways in which military theory, operative concepts, doctrines, assumptions, understandings, and truths are linked to different conceptions of science, technology, war, organization, and ways of imagining warfighting.[55] While military imaginary is a concept that describes collectively held and institutionally and organizationally stabilized versions and shared understandings of what war is and how to prevail in and from it, it is importantly, at the same time, future-oriented visions—descriptive of the world and prescriptive of how to attain this world through the use of armed force. Imaginaries thereby assign logic and legitimacy to the role of armed force in the larger security environment, and how they look into the future.

Targeting exists within these military imaginaries about war as they give rise to certain martial epistemologies, and it is from these that the martial quest to know and operationalize the enemy emerges. It is in this sense that we can say that targeting, as a martial epistemology, orders and configures heterogenous elements according to military imaginaries. Imaginaries are therefore fundamental to targeting's operative logic and ordering work.

Although imaginaries are a condition for the coming together of things, they are also a consequence of targeting—being shaped, altered, and transformed in and through practice. The important point here is to move beyond the notion of imaginaries as static collective systems of meaning that give rise to certain things, be they communities, technologies, or future orders, and to also account for the ways in which imaginaries are constructed, altered, and transformed by the very same forces. While targeting may overall be directed and conceived of through specific military imaginaries and political contexts, in its operation it carves out its own space as much as it is guided by particular orders, intentions, and wills.[56]

Importantly, however, we cannot focus only on military imaginaries here, as the epistemic operations of targeting consist of multiple overlapping imaginaries that in various ways shape the coming together of military targeting's configurations and how they operate. Of particular importance here are the more subtle and often invisible assumptions, values, and biases that inform design choices, data production, data processing, and modeling that do not necessarily stem from the military but are part of the broader martial apparatus. As such, I am interested not only in how assumptions, values, and biases are inscribed into specific configurations but also in how these assumptions are molded into those that seek specific targeting configurations as solutions to their problem.

While configuration, like assemblage or apparatus, emphasizes the bringing together of things, techniques, and practices, its particular attention to how imaginaries and materialities are joined together brings a different focus that can help highlight the inside of a sociotechnical system and its constitutive outside, the histories and encounters through which things over time are configured into meaningful existence, but which also through these processes largely become invisible.

Military Targeting as Epistemology

This also paves the way for thinking about military targeting not simply as an arrangement of things but also, and importantly, as a force of its own—a relational process, working as a particular operative logic, a hinge, a mode of ordering, or framework, through which various imaginaries, technologies, and practices are designed, engineered, and programmed together, forming specific configurations that tie humans and machines to particular operations for the production and operationalization of the enemy.

Focusing on engineering and design choices means that one must locate the circumstantial intent, the functionality, or a specific imaginary that is built into sociotechnical systems, engendering specific human-machine relations and configurations. It might be that acting in the world, or affecting outcomes, does not require intentionality as Bruno Latour would argue,[57] but like Suchman, I want to capture human intent and agency because they matter for what the configurations can be and for structuring the ways in which they operate in and through the world.

While this might open the book to critique for resorting to substantialist arguments on behalf of the designers and engineers, treating them as central connectors, these do not stand outside of larger assemblages untainted by either cultural imaginaries, existing technologies, or the promises of the system they are about to design. Humans and their intentions are not situated in a vacuum unadulterated from their ecological surroundings. On the contrary, meaning and intentions are interactionally contingent and deeply embedded in the various relations they have to other things.

The point of departure here is thus that not only humans have agency, but a great many things in this world also have agency. This is an assumption that, while rocking the boat of much traditional scholarship, has become standard in many disciplines, especially those interested in the role of matter and technologies in this world. Extending agency to technologies means that they are more than neutral tools or instruments extending human will. They alter the horizons of what can be known and acted upon, shaping, enabling, or determining certain options while closing off others. By structuring human relations with the world, technologies have profound consequences for social, cultural, and political practices.[58] As such, it is important here to be wary of how material objects or technologies may shape or alter intents.

In line with process thinking and the relational approaches outlined earlier, Suchman not only wants us to focus on particular design and engineering practices but also, and importantly, wants us to shift the focus away from the intrinsic capacities of an entity, human or nonhuman, toward "the capacities for action that arise out of particular sociotechnical systems."[59] Agency from this perspective is not held by or an inherent property of individuals or things but, rather, is an effect of their relations, emerging from historically engendered configurations.[60]

A commitment to agency as a relational effect in which all things are constantly exchanging, influencing, shaping, and working inseparably does not deny human agency. For instance, there is, of course, no technical agency without humans who fund, design, build, power, maintain, and repair it, but also, humans do not stand outside of the technical or the natural environment it is embedded and embodied in. There is, as such, not any human free will in the liberal humanist sense.

Treating agency as relational thus stands in contrast to Nordin and Öberg, who argued that human agency disappears and dissolves in the process of military targeting, in which the process itself becomes the agent.[61] Anna Danielsson and Kristin Ljungkvist, in their study of military targeting, critique Nordin and Öberg, arguing that although human agency has become increasingly diffuse and hard to nail down in these processes, it has not disappeared. Their ethical project, in contrast, is to recenter and locate human agency in targeting processes, which they argue is essential for critical scholarship on war and for questions about accountability in military operations.[62]

Approaching agency as relational and seeking to study targeting's epistemic operations, I argue that we should not only seek out human agency, separating and disentangling agency and sorting it into "kinds," differentiating the human from the technical or the technical from nature. This is not to deny human agency (on the contrary, humans are involved everywhere along military targeting's operations) but, rather, to accept that agency is a relational effect of particular configurations. From this point of view, human agency does not dissolve but, rather, becomes distributed and reappears in different spaces and temporalities across many different operations.

As such, focus should be on locating novel forms of agency that emerge through different configurations and relational entanglements. It is namely these that give rise to established practices that control and regulate how bodies and machines operate intentionally. But we must also not lose track of how this generates "patterns of unintentional coordination"[63] and what the effects of these. Key to this argument is not only that changes to the configurations of elements in targeting systems and practices produce different models of the enemy, but also that these relational entanglements defy deterministic and predictable trajectories because of their repetitive operations in the systemic encounter with the environment in which they are placed.

Military Targeting as Epistemology

While configurations form the basis for knowledge practices and knowledge claims in which ontologies are concomitant with epistemologies about what constitutes evidence and important knowledge, their many epistemic operations also shape and alter the very foundations of the martial practice of targeting. These are dynamics that are important to capture.

Significantly, by way of Lucy Suchman and Jutta Weber's work on human-machine autonomies, configurations create or alter not only the very making-up of the targets and enemies that targeting produces but also, and importantly, military targeting's "operators."[64] Humans are not only designers and engineers, and machines not only guide human cognition through the designed and engineered infrastructures, but they are both in process of becoming with and through the enemy that is constantly reproduced.

Looking into configurations historically means mapping military targeting over time to unpack how new forms of agency, power, and effects emerge relationally and contingently according to how elements get (re)configured in different contextual periods and how this engenders different ways of processing information and acting. However, this is not simply about the decision to launch a deadly attack, press the button, or fire a weapon; it is about the larger messy operations of making war possible, in which choices, judgement calls, and decisions are part of a broader mesh of prior choices, judgements, and decisions, all relationally entangled.

A Guide to End-to-End Operations

Before we get into the nitty-gritty details of military targeting at four historical junctures, I would like to take you through one last hurdle—how this project is operationalized. Adding a few words on the methodological approach that build on the many themes already mentioned, I now offer an operational guide that is important to remember throughout the book. This guide is not a recipe or a simple set of instructions that if followed would lead to certain results. It is, first and foremost, a map that guides the empirical journey into military targeting to capture the heterogeneity, messiness, and complexity of martial epistemologies and its operations.

In line with the martial empiricist approach described earlier, this is, above all, an empirically motivated guide. It is oriented not toward creating scientific facts but, rather, toward identifying and analyzing the contingent emergence of recurrent processes and practices, the mechanisms, and the human-machine configurations through which they operate and how they produce specific outcomes. This is a loosely designed approach relying on mobilizing concepts, vocabularies, theories, methodologies, and methods from a variety of fields ranging from critical security, military, and war studies to media theory and from critical geography to science and technology studies and feminist technoscience, to mention a few.

Not only are war and military targeting phenomena that largely defy disciplinary shoehorning, but to study these, we need a transdisciplinary orientation that is adept regarding the martial empiricist and onto-epistemological frame of investigation and is sensitive to relational entanglements. Moreover, to explore the ways in which military targeting pieces together its mosaics of the world and to account for the "end-to-end"[65] apparatus of military targeting—the sequence of operations along which the enemy is made known, targets are identified and evaluated, decisions are made, and destructive forces are unleashed—multiple literacies are needed.

As such, the guide on offer here designs a framework for maneuvering and traversing spatiotemporal scales[66] between the so-called macro-, meso-, and microlevels of military targeting: military imaginaries and martial epistemologies struggling to make sense of the world, war, and the enemy; the design of particular configurations into which imaginaries are engineered and through which martial epistemologies are made actionable; and the ways in which these configurations operate in the environment to process and interpret data and connect it with meaning.

It is an approach that requires us to simultaneously trace the scaling effects of politics and military imaginaries at the same time as paying attention to the "micropractices" and to be able to analytically maneuver across multiple spatial and temporal scales—from sensors to global operations and from the split second of the "whack" to the long duration of history. This end-to-end approach to uncover and reveal war's central operations requires scholars to develop multiple literacies, to become "technicians" as Gregoire Chamayou called for—not so much for grasping how devices, techniques, and technologies actually work, but for how it matters that they work in the way they do.[67]

Moving along the long horizontally end-to-end plane of military targeting from politics and imaginaries to configurations and the "whack," we find a number of epistemic operations and suboperations. *Operations*, *operational*, and *operationalize* are terms that have already been used widely throughout these two opening chapters, but their uses are quite specific. The usage of *operations* here is not an extension of what the military calls the operational level, military operations, or operational art but, rather, is a methodological hinge that can facilitate discussions of situated sociotechnical practices with particular epistemological forces to better understand how knowledge is produced through multiple operations that are linked to a long line of imaginaries, institutional and epistemological assumptions and uses, and configurations and how this knowledge is, in turn, made operative.[68]

Drawing on recent work on the operational in media studies,[69] the aim here is to treat targeting and knowledge production in operative ways. Arguing that knowledge is not simply representational of a world out there, or simply performative, but operative, moves the discussion away from what is represented and how this is performed to what is being made/materialized and what this activates. They key here is that by putting operations at the center of epistemology, we also move beyond the standard frames of performativity in which thought and language are treated as the focal point of knowledge. *Operational* is a term that connects us to cognition and the sociotechnical side of knowledge production.[70] As such, it brings us into the gap between the epistemological and the ontological, between how we know and what we know.

The point of departure for this end-to-end approach, in which the goal is to traverse the spatiotemporal scales between the so-called macro, meso, and micro of military targeting, is the functional tasks to be performed to solve key analytical challenges and problems that the martial apparatus supposed at different historical junctures. More precisely, this means starting from historical targeting methodologies, like the joint targeting cycle (figure 2.1). Because targeting methodologies function as coordinating infrastructures, binding together the series of operations or operative moments that are executed at different times and localities, this is a good starting point for following the targeting processes chronologically and tracing the effects of the specifically situated sociotechnical practices.

The aim is then to follow the targeting process along its horizontal plane of operations and then zoom in and drill vertically into the specificities of its operations and suboperations. This deep drilling along a vertical plane is about immersing ourselves in heterogeneity to get our abstract hands dirty by attending to the tiniest force of detail,[71] showing that the very structured and gradualist depiction of knowledge production is, in practice, messy and emergent rather than the straightforward piecing together of different elements. But, importantly, we have to do this without losing track of the sociopolitical and historical embeddedness and constitution of these practices by always zooming out on the imaginaries, politics, power, and devastating effects of military targeting.[72]

Traversing scales, horizontally and vertically, and zooming in and out, offers a dynamic approach that can bring out the particular operative moments and the multiple operative scales that play out simultaneously, teasing out the open-ended processes of becoming, the distributed and relational aspects of agency, and can show how violence and harm are extended into different spaces and temporalities beyond the immediate destruction on the ground.

Although what is ultimately at stake is the various ways in which violence operates in our world and its devastating effects on people, I am not predominantly focusing on the death and destruction on the ground but, rather, on its conditions of possibility. Thus, the empirical targeting "episodes" and examples in these pages are not explicitly violent but, rather, are attuned to specific epistemic processes and technical operations at different scales. This is not to shy away from the terrible violence targeting brings but to emphasize different aspects of targeting and to compliment other investigations bearing witness to the horrible effects of war. To show how violence is embedded in epistemic operations is a way to connect to the critical approach in this book of extending our analysis of violence into different spaces and temporalities.

While this approach is similar to the preferred Science and Technology Studies (STS) method of "follow the actors"[73] to map the route of material objects of science, the end-to-end approach here is rather to "follow the operational" in epistemic operations—zooming in and out on the operations needed to configure systems, produce and process the data, and interpret the outputs and the politics and violent effects of these operations. This end-to-end approach is thus a tactic that addresses the nitty-gritty workings of epistemic operations and their various techniques, which, at the same time, does not reduce these to mere technical operations by always being cognizant of their sociopolitical situatedness, imaginaries, and histories and their devastating effects.

There is, however, a caveat to this "targeting in action" approach. The favored ethnographical methods of STS do not immediately offer themselves as a solution to issues like war and martial practices primarily because these are practices shrouded in secrecy, but also because there is no single STS "laboratory" that we can replace with the complex, multisited, and multitemporal epistemic operations of military targeting. Although this project cannot follow in the same manner, what I take away from these approaches is the need to account for the spaces between discourses and practices and develop an understanding of how knowledge is produced in and through sociotechnical practices and operations. This is not about exposing secrets, but about "navigating secrecy," highlighting what can be known and to some extent inferred from mapping a variety of sources.[74] By mapping, I mean assembling an open, nonexhaustive archive of bits and pieces, collaged together, allowing for multiple entryways for new data and insights into lines of flight.[75]

Military Targeting as Epistemology

The strength of my argument in this book depends on using multiple methods to collect and analyze data from a variety of sources with the added ingredient of speculation. This is especially significant with regard to the present and the future. As Jairus Grove has argued, "Without the blueprints, we have to creatively speculate about the conjunction of heterogeneous actors."[76] Even if the blueprints were available, what work this targeting assemblage would do would not be certain, nor for that matter would be the processes of inclusion or exclusion of certain elements or how the assemblage would transform or change. Thus, this research relies on what can be inferred from different military imaginaries, rationalities and data logics, technologies, and practices and how these are held together—the operations that sustain them to make targeting and its effects possible.

Not offering a solution to the complex issues under the microscope but, rather, accepting that different stories and narratives could have been told, had other things and materials been collected, juxtapositioned, and layered, I do attempt to reduce the risks of shoehorning complex phenomena and processes into overly simplified registers. Although this more experimental practice may come at the expense of precision, it provides for greater flexibility in the research, allowing the unexpected to be followed, detecting things, issues, and relations that might have been left undiscovered had I followed a more traditional way of linking methodology to methods and applying prescribed techniques. In this way, speculation becomes part of the process, as the research is trying to let empirics and theory come together to create an analytical story of military targeting.[77]

Toward Ecologies of Operations

If one takes the systematic and operational interactions between the targeting configurations and the environment, between the enemy modeled and the enemy "out there," seriously, it is perhaps better to refer to these martial relations in ecological terms. Although *ecology* is a term that often publicly and in scholarly discourse has come to mean the complex entanglements of human and nonhuman processes and life forms, it often suffers from an "organicist" bias that excludes the built world and the power relations inscribed in it. In contrast, ecology here centers its gaze on the ways in which sociotechnical configurations and their operations engender particular "ecologies of operations"[78] in their systematic interactions with the environment.

If we follow the notion of operations here, we can talk about an operational-ecological notion of relationality that, in contrast to the more "connectionist" notion networks and systems, treats the associated environment as a recurrent regime of reciprocal causalities, in which relations take the form of operational coupling.[79] Ecology is thus a trope that focuses less on the links and the distribution of agency within the network and more on how different elements effect and affect each other and are, as such, more attuned to what Leigh Star argued was the countless spaces left as background by network thinking.

Allowing us to look for the "spaces between,"[80] or in the words of Maria Puig de la Bellacasa, to the sites of "intensities, synergies and symbiotic processes within relational compounds,"[81] thinking in ecological terms creates the possibility to treat the intimate relations and systematic interactions between the war machine and the environment as an ecology of operations and how these function in cyclical symbiotic ways—always in in transition, constantly rearranging connections and relations. If, as Hayles argues, interaction between configuration and the environment is key to generating context and meaning-making, the dynamic epistemic operations do not only respond to their environment in ways determined by their internal self-organization. It is not simply reproducing what they are designed to see, and know what they are intended to know, but is emergent.

Here, it is useful to think about Brian Massumi's distinction between systems and processes. Systems, he argues, desire the reproduction of their forms and functioning, while processes are tendencies in which repetition produces difference and, thus, becoming. While systems are expressions of martial imaginaries and operative processual logics concretized in specific configurations, the interactions through which these configurations encounter the environment generate novel capacities to affect and be affected, engendering different relational entanglements.[82] Military targeting can thus be said to produce worlds, which targeting not only comes to depend on for its operations, but which is also a condition for its development and transformations.[83]

Thus, we can think of targeting configuration's repetitive epistemic operations as tendencies that produce difference and thus also transform in unexpected ways. It might be that targeting works as an operative logic that combines an ontology with an epistemology in autopoietic ways, endowing itself with the powers of self-causation, but its systemic interactions with the environment mean that it is emergent. In these dynamic operations and encounters, meaning develops in context, and the choices mutate and adapt as the various micro-operations of targeting produce and operationalize worlds. As such, the notion of operational ecology inserts temporalities into spatial structures, sensitizing the analysis to timing and processes as well as structure and architecture. It thus helps to contextualize the meaning-making practices of the configurations, seeing how both the modeling operations and the actions engendered encounter the enemy object.

In this worldview, the enemy and its targets are not only inanimate objects of warfare but active participants and components in the (re)production of this worldview choice and decision. An object that that is not only unruly, actively seeking not to be captured by the processes of the targeting configurations, but is in many ways also engaged in modeling itself, seeking to disrupt, degrade, or otherwise outwit the targeting assemblage. The enemy is, in the words of Norbert Wiener, not the "'the Augustinian Devil"–likened with nature–which might try to resist your scientific inquiries, but "the Manichean Devil" deliberately trying to trick or mislead you.[84]

Adding agency and mobility to epistemic objects, showing that these are not passive and discrete things but processes that play an active role in the production of knowledge, ecology places emphasis on the in-between of the abstract reality that is captured and the concrete actualizations of targeting that take place in various operational environments. For these reasons, the knowledge, choices, and decisions produced through military targeting cannot only be analyzed as establishment of facts, boundaries, or categories according to some predefined strategic rationality, but must also be viewed through the actions they generate, their operational goals and effects and how these form particular ecologies of operation. It is, in other words, necessary to understand the translation between targeting in the abstract, as described and prescribed by doctrines and handbooks, and the concrete actualizations of the ecology of operations, both in its cognitive processes of modeling and on the ground as in military operations.

Making the relationships between military imaginaries, human-machine configurations, and their systematic encounters with the environment the focal point of the analysis, this book offers a way of thinking and investigating multiple processes and scales simultaneously that, at the same time, foregrounds the world-making capacities of epistemic operations. It is not a quest to find their true or hidden origins but, rather, a means to explore their sociotechnical and historical complexities that activate multiple temporalities of the present operations and future ramifications of military targeting.

By investigating targeting's world-making capacities in detail and showing what they do, we can begin to question the epistemological systems underpinning the practices of military targeting, the worlds they produce, and the violence they engender. The aim is to provide a more nuanced understanding of its formative forces enclosing the world in particular imaginaries, and epistemic practices and infrastructures, and chart possible alternative imaginaries and future beginnings. These are, however, empirical matters that the following pages will explore.

3

The Enemy as a System: The Geopolitics of Strategic Paralysis

In essence, air power is targeting, targeting is intelligence and intelligence is analyzing the effects of air operation.[1]
– Colonel Phillip S. Meilinger, United States Air Force

It is general consensus among military professionals that the art of targeting, in its modern conception, was born and developed alongside air warfare. This is not only because targeting is considered the essence of air power, as the opening quote suggests, but because with the invention of aviation a new spatial dimension opened, leading to new ideas about strategy, tactics, governance, and control and how to achieve this.

The question of who, what, and how one targets has, of course, been part of human reasoning since the invention of war,[2] but air power profoundly changed how the enemy was perceived and thought about. Although balloons and zeppelins had previously been utilized, with limited success, as vehicles for reconnaissance and bombing, wars had largely been fought on land and at sea.[3] Up until and including WWI, ground combat operations, techniques, and materiel were largely designed to seek out and engage enemy forces directly on the battlefield, to attack fortifications or logistical facilities, or interdict supply routes. In large part, war was about seizing territory from opposing enemy forces through battle, holding them, and forcing the enemy to surrender.

The concept of the battlefield thus structured the understanding of war as something that occurs within a limited space and time in which the course of war is settled. As such, the battlefield served as an important construction of reality—providing logistical, psychological, legal, and normative constraints upon war—an ideal through which militaries could design and engineer their approaches to warfare.[4] This ideally constructed space-time would, however, gradually be reshaped and reconfigured by developments in air warfare, constructing novel "aerial gaze" mechanisms through which the enemy increasingly came to be seen.[5]

Opening up a new geographical dimension to warfare, the airplane made it possible to target not only enemy military forces on the immediate battlefield from the air but also, in theory, the whole socio-economic system and military-industrial complex of the enemy, largely bypassing its ground combat forces. However, determining how the enemy system functioned, its vulnerabilities, and how to exploit these, was not straightforward and became an increasingly vital preoccupation for military planners seeking victory from the air.

Although airpower and the aerial view are crucial to the emergence and transformation of military targeting, this chapter, in line with the overall text, tells a different story, a story which intermingles with the rise of what Hunter Heyck has called the "age of system"—a specific way of seeing the world as systems with its own way of understanding the world, producing a particular kind of knowledge about the function of modern societies.[6] Akin to Paul N. Edwards' query, "How did "the world" become a *system*?",[7] at the start of his account of climate science as a "global knowledge infrastructure," this chapter is an exploration into *how the enemy became a system*,[8] and with what destructive effects.

By the 1930s a number of technoscientific trajectories were converging with military thought and practice to enable a doctrinal reimagination of war and the enemy and give rise to what would be known as systems warfare. Transforming not only the spatial and temporal rhythms of warfare, but in large part what the enemy was, this new martial epistemology produced an understanding of the enemy as an industrial system composed of causally interlinked nodes that, if effectively destroyed from the air, would lead to the certain material and moral collapse of an enemy.

The cementing of targeting as a special discipline during WWII made sure that in the "industrial age" of total war the enemy was no longer simply a collection of armed forces but a collection of critical infrastructural nodes. Constructing vast archives of targets and matching them to desired end states, the military rearranged the world with devastating effects, turning cities and urban centers into battlefields, and warfare into massive slaughter of civilians.

The Airplane, Strategic Bombing, and the Birth of Modern Targeting

When WWI broke out, aviation was a novel technology and the martial apparatus lacked concepts and doctrines for its use. Although airplanes provided militaries with the capacity to target elements beyond the immediate frontline, during the war, aircrafts were mostly used for reconnaissance, to photograph the terrain and map out the frontlines, forts, and trenches and to fine-tune and pinpoint locations of enemy forces, providing intelligence for ground artillery bombardment.[9] Tactical uses of aviation—interdiction and close air support—were by and large invented and improvised as the war went on, and strategic bombing during the war was, as Thomas Hippler argues, hopelessly disconnected from military, technological, and industrial reality.[10] While this resulted in very limited effects on the course of the war, its effects on the conceptual role of air power and strategic targeting were remarkable.

WWI saw the expansion of specialized aircraft for bombing and showed the potential for the so-called concept of rear-area strategic bombing. During the war strategic bombardment plans were made to bomb commercial centers and lines of communication (LOCs) to cut off the supply chains to armies in the field. To do this, planners were put to work to determine critical enemy industrial centers and LOCs to create targets for air operations. Though these plans would not be set in motion before the war had ended, they signaled the birth of strategic air targeting.

After the war, the US Air Service conducted a so-called bombing survey to investigate the effects of airpower during the war to improve its use. This postwar survey was to influence and shape bombing operations and policies in the years to come and served as the seed to what was to become the martial discipline of military targeting. One of the main conclusions from the WWI US Bombing Survey, based on British and to some extent French bombing during the war, as the Americans did not have enough aviation operations to warrant investigation, was the need to identify critical targets to support a systematic plan for air operations. The survey stated,

> The greatest criticism to be brought against aerial bombardment . . . as carried out in the war of 1914–1918 is the lack of a predetermined program carefully calculated to destroy by successive raids those industries most vital in maintaining Germany's fighting force. The evidence of this, is seen in the wide area over which the bombing took place as well as the failure of crippling, beyond a limited extent, any one factory or industry . . . the bombing of a town rather than some definite objective of military value in the town . . . the wide spread of bombs over a town rather than their concentration on a factory, is not a productive means of bombing.[11]

Rather than seeing the disconnect, or the apparent limits of air power, the US Bombing Survey lamented the implementation of air power in WWI and identified the lack of intelligence as the weakest link. Under the headline "Suggestions for Future Bombing Campaigns," the survey recommended,

> The three kinds of bombing that are of most importance are, first, that directed against war industries; second, that against railroad lines; and

third, that against an enemy's troops in the field. In considering the first a careful study should be made of the different kinds of industries and the different factories of each. This study should ascertain how one industry is dependent on another and what the most important factories of each are. A decision should be reached as to just *what factories if destroyed would do the greatest damage to the enemy's military organization as a whole.*[12]

In this short passage, the seeds of what would be known as the discipline of military targeting were sown. It was based on a systems logic suggesting that neither supply lines nor enemy troops are of vital importance. Rather, one should study the enemy's military-industrial complex and determine its relational interdependencies and from that decide which factories, if destroyed, would do the most damage to the enemy's ability to wage war.

The survey also concluded that wide area bombings of towns neither affected the morale of the population nor effectively hampered the enemy's ability to sustain the war effort. Champions of strategic bombing thus did not object to the fact that strategic bombing had proved to be ineffective during the war but argued that the meager effects were the result of inadequate implementation, not of the ideas. It was an epistemological argument. If only we had better knowledge, then....

As air power technology evolved, so too did the concept of strategic targeting and, with it, the debates and theories concerning the logic, rationale, and legitimacy for selecting specific targets for the purpose of winning wars. Famous air power theorists in the interwar years all argued that strategic bombing would avoid the terrible trench-style attrition warfare that WWI had become, because it offered a method of waging war beyond the attrition-based model of ground warfare. As prominent British military theorist Liddle Heart argued in 1925, "Aircraft enables us *to jump over* the army which shields the enemy government, industry, and people, and *so strike direct and immediately at the seat of the opposing will and policy.* A nation's nerve system, no longer covered by the flesh of its troops, is now laid bare to attack."[13]

Armed with optimism about the future prospective for air power, its proponents argued for its revolutionary potential, believing that air power could destroy the enemy's centers of gravity and achieve victory. Hitting industry, communications, government, and the population at large, they hoped and argued, would see to it that governments would succumb to the pressure and surrender. However, wanting to avoid the slaughter of the trenches, the carnage would now move to urban civilian centers.

To support "precision bombing," the Survey had concluded, an organization with a constant focus on systems analysis to produce targets and maintain files of information about potential targets as well as requisite target materials would be required. As such, there was a need for officers "to compile and maintain all information of value in the preparation of bombing missions, an indexed file of photographs, and a stock of maps and charts showing bombing targets and intelligence concerning them."[14] But first and foremost, what was needed was to produce information about the interdependencies of the military-industrial complex and the most vital substrates of that system. What was needed was a systematic process of producing and operationalizing the enemy, namely, military targeting.

The Industrial Web Theory

While the development of strategic bombing theory was well under way in the Euro-Atlantic sphere, a parallel, if not separate, form of air warfare was ongoing in the colonial peripheries from Ireland to Africa and India to the Middle East—that of colonial air policing. According to Thomas Hippler, the history of air warfare has almost exclusively focused on strategic bombing in World War II and during the Cold War and afterwards, leaving out the importance of its colonial precedent, where bombing was experimented with and perfected[15] to the point that it became a universal feature of colonial domination[16] and a key mechanism of order-building.[17]

Notwithstanding the important history of colonial air policing for the development of air power, the ideas and theories of strategic bombing and targeting derived their imaginaries, epistemologies, and practices not from the experimental colonial laboratories but at the office desks of European and American air forces. While these parallel developments certainly overlapped and influenced each other at times, in particular, in their discursive and narrative production of Others to justify and legitimize bombing, strategic bombing derived its theories and practices not primarily from experiences in war but from deterministic Newtonian mindsets and techniques that saw modern industrial societies as machines, webs or systems that depended on effective links between critical nodes. This was, as Hunter Heyck has termed it, "the age of system."

Between 1920 and 1970, Hunter Heyck contends, "virtually every field of social science reconceptualized its central object of study as a system defined and given structure by a set of processes, mechanisms, or relationships."[18] This type of thinking would lead to a particular way of studying objects with an "idealization of formal, instrumental reason, epitomized by the development of systematic theory and formal modeling, and by a fascination with procedural logic, as in the development of algorithms, heuristics, protocols, decision rules, production systems, and programs."[19] Emphasizing the "behavioral-functional analysis of organizational or structural-relational properties," systematic theory would create conceptual representations of some aspect of reality, or models.[20]

Seeing the enemy through such lenses, it was believed that empirical observations could be turned into consistent relationships among certain sets of variables. Relationships or dependencies, strategic bombing theorists argued, created vulnerabilities that could be exploited by novel air power, to induce systemic collapse or paralysis of an enemy's war-fighting capabilities and will. Military targeting was not interested in governing populations but, rather, sought the most effective way of destroying a modern industrial enemy. Colonial air policing and strategic targeting were different epistemologies and required different types of knowledge.

While air policing doctrine and tactics were perfected in the colonial peripheries, much intellectual effort went into debating and developing the ideas and practices of strategic bombing. Although the proponents of strategic bombing agreed that bombers could now quickly destroy a nation from the inside out rather than by slowly defeating an enemy from the outside in, as had been the practice for centuries, the debates on where to focus the effort raged on. Despite the conclusions from the US WWI bombing survey that the future of strategic bombing had to be conducted through "precision strikes" at selected industrial targets, other influential individuals continued to believe in targeting the morale of the population.

In his seminal 1921 book, *Command of the Air*, Italian strategist Giulio Douhet argued that air forces had quickly become the primary means for conducting war and that armed conflict could be rapidly won through the implementation of a strategic air campaign directed primarily against the enemy's civilian population. A successful aerial bombardment would demoralize them to such an extent, the argument went, that they would revolt and force their government to seek peace.[21]

However, morale, the will to fight, or popular support were, in the eyes of the faculty members of US Army's Air Corps Tactical School (ACTS), all incalculable objects based on "unscientific" assumptions about human collective psychology. Rather, they argued, following the recommendations from the WWI bombing survey, focus should be placed on paralyzing the enemy through targeting its industrial capacity to wage war. In the text entitled "Employment of Combined Air Force," a series of working propositions established an intellectual foundation for the work of the ACTS:

"Proper selection of vital targets in the industrial/economic/social structure of a modern industrialized nation, and their subsequent destruction by air attack, can lead to fatal weakening of an industrialized enemy nation and to victory through airpower."[22]

By the 1930s, these propositions eventually led to the development of a specific and unique air doctrine alongside a belief that air power could achieve victory. The doctrine, better known as the *industrial web theory*, hypothesized that the destruction of one critical stress point would create a choke point that would severely limit or curtail the enemy's war-making capacity and eliminate the need to take out the enemy's entire industrial output or harm civilians. In effect, the doctrine assumed that the more effectively industrialized the enemy was, the more vulnerable it was to strategic bombing. As US Air Force Major General, Haywood S. Hansell later wrote, the ACTS argued for "1. Destroying organic industrial systems in the enemy interior that provided for the enemy's armed forces in the field: 2. paralyzing the organic industrial, economic, and civic systems that maintained the life of the enemy nation itself."[23]

Although this was also based on a set of assumptions about how modern societies functioned, it was nevertheless, for the ACTS, assumptions that were calculable. The ACTS theory, in contrast to the moral centered approach of Douhet and others, supposed a predictable and calculable cause-and-effect relationship between bombs and effects. Through "scientific" rational calculation, it was believed, one would be able to determine a set of targets that would trigger the mechanism that would lead to the predicted military and political end. This paved the way for thinking about targeting in different ways.

While ACTS emphasized the destruction of the military industrial complex (i.e., the capability to wage war), they also stressed the ability of targeting to destroy the will and morale of the enemy through bombing the "vital elements upon which modern social life is dependent."[24] Understanding the state as a complex system of economic and social dependencies, as "factories," it was possible to conjure new ways of inducing the collapse of the system's capabilities *and* the will to fight.[25] They had conceptualized, in the words of Tami Davis Biddle, "an aerial shortcut to the accomplishment of a Clausewitzian dictum: to end a war one must destroy the enemy's will to fight."[26] Such cause-and effect models, however, depend on accurate intelligence on potential enemies and their vulnerabilities and precise locations and data on targets. Thereby, the ACTS concluded, in the interwar years, that strategic intelligence on the major powers of the world was "vital to planning and operations of strategic air warfare."[27]

Seeking to alleviate the horrors of WWI, strategic targeting theories and their systematization of the enemy would create new and more devastating horrors. Viewing the enemy through this novel military targeting epistemology effectively reconstructed the battlefield and expanded and subsumed the enemy into a collection of fixed industrial targets and urban centers, seeing to it that there was no longer any viable distinction between military and civilian. As J. F. C. Fuller, another famous British war theorist, contended, "In modern warfare it is pure sophistry to attempt to draw a line between those who fight and those who assist the fighter, since entire nations go to war."[28]

Seeing war as total and civilian populations as complicit in the war effort, and not wanting to repeat the slaughter of soldiers during WWI, military theorists did not have much difficulty in arguing for the strategic bombing of entire nations, shifting the slaughter from the muddy trenches to the flammable interiors of urban centers. This normative and technopolitical discourse, along with the dehumanization, demonization, and racializing of the Other, established the conditions of permissibility and possibility for the practices of military targeting to produce an enemy figure that would bring the war directly to the civilian population.

A Newborn Science

When World War II started, the US Army Air Corps had a mature doctrine of how to employ its strategic bombing theories. However, it lacked the intelligence on targets needed to operationalize its doctrine of paralyzation, not to mention the almost complete lack of empirical data on the effects of bombing.

To amend the lack of targeting intelligence, General H. H. Arnold established an air intelligence organization under the chief of the US Air Corps in 1940. The Strategic Air Intelligence Section, A-2, was tasked with undertaking economic-industrial-social analysis, to investigate and describe vital and vulnerable systems and select targets and prepare target folders. However, when the Americans entered the war, the air force still lacked the intelligence needed to plan and conduct combat operations and a systematic method for selecting targets. Despite the recommendations from the WWI bombing survey, John Glock states, "There was still no organization capable of doing the systemic analysis required for proper targeting. There were no trained target intelligence officers [and more importantly, no] data base of potential targets to build the target material needed to turn the industrial web theory into action."[29]

While the ACTS attempted to develop a model of modern industrial society, conceptualizing it as a complex system of interdependent subsystems, alongside a theory on how to organize its demodernization through aerial bombardment,[30] the lack of proper targeting analysis was problematic. By the end of 1942 the Army Air Force (AAF) had accumulated vast amounts of data on the German military-industrial complex, but still, "no rational system for target selection existed."[31] Data, it turned out, did not in themselves provide knowledge, understanding, and decisions without proper analysis.

To overcome this shortfall, the US created the Enemy Objective Unit (EOU) and the Committee of Operations Analysts (COA), which later formed the first Joint Targeting Group (JTG), responsible for the collection and analysis of intelligence for the purpose of air target selection. The COA was what we today would refer to as interdisciplinary, consisting of a diverse group of people (military personnel, engineers, economists, lawyers, and so on) tasked with providing a comprehensive analysis of the German war economy to guide the bomber offensive. They were specifically looking for war-making capabilities and, through *operations analysis*, developed criteria for first-order and second-order success (ripple effects) rates for bombing.[32] The use of operations analysis was key, as it was a methodology that allowed analysts to focus not only on the targets themselves but on how specific targets were relationally linked in a system. Operations analysis, according to Peter Galison "was essentially a methodological theoretical reconstruction of the interconnections that held together the German economy and war machine and that asked how it could be blown apart." In effect, he argues, it created and reproduced a "technocratic vision of the technical enemy other."[33]

Although the theories and hypotheses of strategic bombing and targeting were well established during WWII, techniques and procedures for how to conduct target analysis were still in their infancy. The British conducted their own strategic inquiry based on German bombing of their own cities, factories, and ports. Damaged areas were examined for economic and structural susceptibility to various types of bombs, and as time went on, a pattern developed that would serve as the theoretical foundation for techniques and procedures for target analysis in order to attack the German-held European continent.

This study, known as the Princes Risborough, named from the town where the analysts had gathered, drew on a number of civilians—"architects, civil, industrial, chemical, and fire insurance engineers, mathematicians and economists"—to help with the "new born science" of target analysis.[34] These studies, which also included military personnel from both sides of the Atlantic, helped to cement the view of strategic bombing as the "long range attempt to destroy the will and ability to make war by attacks on industrial and civilian economies."[35] The new science of target analysis would work as an intelligence process with distinct operations, beginning with an examination of those items that must receive attention according to the logic.

Having selected targets, target systems—a collection of various elements within a category such as electricity or oil—emerged and evolved according to their importance to the war economy. Once target systems had been produced, an appraisal of the system's vulnerability in general and of the individual targets within the system was made according to the desired disruption in order to determine the type of attack necessary to produce this effect. Targets and target systems were in the end selected by "calculating the probable returns from systematic attack on alternative target systems against the comparable costs."[36] Taking all this into account, the EOU could now assign value to each target and the various target systems, with little consideration of other factors, such a civilian death tolls, apart from the importance of each target to the whole.

Target analysis naturally rested on the quality and quantity of the information available, and frequently, during the war, this was of limited supply. As such, analysts experimented with assembling and processing information from multiple sources, "pre-war publications, newspapers, trade journals, etc., refugees, prisoners of war, government statistics, captured documents, captured equipment, radio intercepts, and aerial photographs," and putting together "bits and pieces" to "get as complete a story as possible."[37] To supplement the collected bits and pieces, analysts travelled to industrial plants in Britain that were similar to those in Germany in order to gain key insights into how each plant figured in the German industrial system and for information concerning aiming points for each specific target and possible damage assessment.[38]

By 1943, the EOU had crystalized and agreed upon a framework of target theory that focused on the need to locate key nodes that would cause ripple effects throughout the German war machine. The assumption that such key nodes existed led the Americans to focus on seemingly peripheral things such as the German ball bearing industry in the hope that the loss of this key industry would grind the whole war machine to a halt.[39] While finding such key nodes was the dream of target analysts, the importance of such targets rested on their relations to the system as a whole. Following this logic, the analysts concluded that because ball bearings have a relationship to all machines, and thus also to all industry, they were a vital node in the German war machinery.

One of the major problems was that the bombing campaign itself was difficult because so little was known about the effects of large-scale bombing on the enemy's industries, let alone the highly intangible entity of enemy will. There were few precedents for determining and selecting appropriate objectives, targets, and measures of effectiveness besides what could be theorized, hypothesized, and assumed, and the EOU had almost no experience in gathering and analyzing the intelligence necessary for such operations.[40] While it was believed that the infrastructures and the materiel would speak for themselves as long as all the bits and pieces were collected, in the final instance, the decision on which targets to nominate for attack and which ones to prioritize rested on human judgement and "gut feeling."

From the Calculable to the Incalculable

In Britain a similar and parallel approach to the EOU was conducted by the British Ministry of Economic Warfare (MEW). Known as the Bomber's Baedeker—presumably after the German publisher Baedeker, famous for their travel guides—this was a list of German towns to be bombed, selected for its industries and targets deemed most valuable for the German war effort. Serving as an operational guidebook for selecting targets, the Bomber's Baedeker contributed to the spatial distribution of destruction.

Uta Hohn, offering a geographical mapping and reading of these lists shows that there was a staunch disagreement between the RAF's Bomber Command and the MEW. Not only did nocturnal "precision" bombing against small industrial plants prove to be difficult and in large part ineffective,[41] but the Bomber Command, under the weight of its infamous chief, Arthur T. Harris, commonly known as "Bomber" Harris, was "obsessed by the idea of demoralising the civilian population by *indiscriminate* area bombing of major cities and later also medium-sized towns."[42] In two letters to the Chief of the Air Staff, Charles Portal, Harris lamented the efforts of MEW: "In the past MEW experts have never failed to overstate their case on 'panacea', eg ball bearing molybdenum, locomotives, etc., in so far as, after the battle has been joined and the original targets attacked, more and more sources of supply or other factors unpredicted by MEW have become revealed. The oil plan has already displayed similar symptoms . . . I repeat that I have no faith in anything that the MEW says."[43]

Instead, Harris put his faith in destroying the enemy's morale and will to fight through area bombing. To him, demoralizing the population, "Douhet style," would serve this end. Despite the efforts of the MEW, and the EOU, on focusing the bombers on key industrial targets, the sciences and logics of targeting did not win over the deep-seated assumptions and ingrained ways of doing. In January 1943 a compromise was made at a meeting among the Allies in Casablanca: "The strategic air forces was established as the 'destruction and *dislocation of the German Military, industrial, and economic system* and *the undermining of the morale of the German people* to the point where their capacity for armed resistance is fatally weakened.' Specific target systems were named."[44]

However, the bombing of cities was a strategy that satisfied both those in favor of crippling the German war economy and those in favor of demoralizing the population through indiscriminate bombings. If you hit industry, it effected the German war effort; if you missed, it was also good, as you would, in theory, affect the German people living close to industrial facilities. Through this sinister way of killing two birds with one stone, the choice of ordnance became firebombs, and the German cities to be bombed were largely selected due to their flammability.

In this way, the assumed calculability of German war industries was coupled with the incalculability of hope—the hope that morale, the will to fight, and popular support would be affected by strategic bombing. In effect, targeting analysis and faith in morale are different ways of understanding collectives and how they hold together: one based on presumed calculability and linear logics, and the other based on incalculable assumptions about human psychology.[45] The result of coupling these two is that targets are found everywhere, to devastating effects.

> The bombing of German and Japanese cities was conceived in the context of an imaginary of war-fighting in which the political, military and social infrastructures of the enemy were understood to have converged in the figure of the enemy nation. In such an imaginary, no proper distinction is made between combatant and non-combatant. Rather, bombing is oriented towards the annihilation of the enemy as a whole: its war machine as well as the society that enables that war machine both materially and discursively.[46]

The horrific and destructive results of this orientation toward the annihilation of the figure of the enemy nation carried out by the Anglo-American Combined Bomber Offensive (CBO) was unprecedented large-scale destruction of civilian and urban areas. Without taking into account the psychological torment and trauma of being placed under bombardment or witnessing the death and destruction of your habitat and loved ones, the destructive figures from Germany are staggering:

> A revised estimate prepared by the Survey (which is also a minimum) places total casualties for the entire period of the war at 305,000 killed

and 780,000 wounded. More reliable statistics are available on damage to housing. According to these, 485,000 residential buildings were totally destroyed by air attack and 415,000 were heavily damaged, making a total of 20 percent of all dwelling units in Germany. In some 50 cities that were primary targets of the air attack, the proportion of destroyed or heavily damaged dwelling units is about 40 percent. The result of all these attacks was to render homeless some 7,500,000 German civilians.[47]

Similar and even greater amounts of death, destruction, and trauma were the result of the so-called strategic bombing of Japan, even without considering the atomic weapons dropped on Hiroshima and Nagasaki.

... and Back Again

Despite the problems of hitting the selected targets with precision and the terrible devastation on the ground, "a well-disciplined air doctrine had crystallized and had been generally accepted"[48] by the beginning of 1944. However, the problems with target intelligence persisted. During the war, while Allied aircrafts were still burning and pulverizing German and Japanese cities, the US Strategic Bombing Survey was established to analyze the effects of the bombing effort. This was a massive undertaking employing over 1,000 people, including industrial experts.[49] Just as the World War I bombing survey had done, the US Strategic Bombing Surveys (USSBS) emphasized the importance of target selection for the planning and conduct of operations. So did General McDonald, US Air Force (USAF) director of intelligence, arguing that "target intelligence is the basic requirement because a Strategic Air Force is nothing more than a large collection of airplanes unless it has a clear conception of what to use its planes against."[50]

Strategic bombing required not only knowledge of geography and potential targets but knowledge about interdependencies between targets, so called-target systems, and how they would collectively lead to the collapse of the enemy. By examining the interdependencies and vulnerabilities of socioeconomic elements, strategic bombing theory generated a new science and a new kind of knowledge about the enemy. Rather than seeing the enemy as largely a collection of military forces, the "age of systems" generated a new way of understanding the enemy. With industrialization came great industrial strength but also, equally, great vulnerabilities that militaries now sought to exploit through a "total war" mindset fixated on the human-built environment as a system-of-systems open to functional disruption. The result was that the vital elements upon which a modern society depends had become the battlefield, and the objective was the total paralyzation or obliteration of this system from the air.

Regardless of the less than effective outcome of the strategic bombing campaigns against Germany and Japan during WWII, the war had demonstrated to strategic bombing's proponents that the proper selection of vital targets—through rational planning based on the systematic processing of information connected to the overall imaginaries of the enemy as a system—is crucial to the successful application of air power. As such, a novel martial epistemology for the production and operationalization of the enemy had been consolidated. As William Walsh, with assured awe, concluded a few years after the end of the war,

> As the awesome spectacle of destruction witnessed by members of the U.S. Strategic Bombing Survey teams unfolded into an evaluation of the many aspects of the air war, it became apparent that the disintegration of a nation's ability and will to wage war could be planned and effected to a remarkable degree through systematic air attacks on its strategic components. This was the proof of the validity of the strategic concept—a concept so embracing that by it warfare was completely revolutionized and the future course of history dependent upon a completely different set of principles.[51]

The onset of the Cold War would, however, pose new problems for which the targeting assemblage would be reconfigured.

Configuring the Bombing Encyclopedia of the World

By the summer of 1947 it had been "established" that the Soviet Union was the enemy and that the weapon of choice in any future war would be atomic and, later, nuclear. War planners at the infamous US Strategic Air Command (SAC), due to the scarcity of atomic weapons and in conjunction with the vast destructive powers of these weapons, had come to the conclusion that one should emphasize concentrated area attacks on urban centers in what became known as the doctrine of massive retaliation—a conclusion largely based on the view and experience from WWII, in particular from Japan, that destroying civilian, economic and industrial centers would cripple the war effort of any state and thereby accept unconditional surrender.[52] However, the SAC approach differed from those of analysts from target intelligence that worked within the doctrine of precision bombing inherited from the ACTS.

Despite the horrors of Hiroshima and Nagasaki looming large over the US military and political establishment, the atomic bomb, while its destructive potential was certainly acknowledged after the two Japanese cities had been obliterated, did not, in the early years of the Cold War, condition radical changes for targeting analysts in the selection of targets. To them, the destructive potential of weapons did warrant any changes to either the theory of military targeting or the prioritization of targets. The enemy system was the enemy system, regardless of the means by which to destroy this. As such, the atomic bomb was simply incorporated into the same doctrines and theories of strategic bombing.[53] The tragic irony here is that any "precision" use of atomic weapons against industrial targets would ensure the doctrine of massive retaliation.

Building on the beliefs and assumptions and the knowledge and logics developed during WWII, those in targeting intelligence began to further organizational routines and standards, practices, and infrastructures for the production, analysis, and modeling of targets and target sets to deliver predictions to war planners. Targeting and its requirements would therefore continue to serve as a guide and basis for a number of technical and infrastructural developments, always referring back to its logical "needs," both enabling and constraining certain paths the targeting configurations could take.

In order to fix the problems of targeting intelligence and develop a systematic approach to target selection, the US military embarked on a massive undertaking. Known at the time as the Bombing Encyclopedia of the World (BE), this postwar intelligence infrastructure commenced in 1946 with the goal of building a comprehensive database of potential targets around the world. By the 1950s, the encyclopedia amounted to no less than "an alphabetical, numerical and geographical INDEX to all possible targets of significance throughout the world [listing] hundreds of thousands of such targets."[54] The goal was to identify strategic vulnerabilities to attack and to recommend targets for destruction. Like the industrial web theory, this strategic vulnerability analysis was focused on finding the key target systems or nodes that everything else relied on, "the destruction of which would cut across the totality, the entirety of the enemy's ability to defend himself."[55] Providing target recommendations for every country in the world, the Bombing Encyclopedia organized a massive amount of information that, in theory, could be rapidly queried to identify air targets in the event of future war.

The task of compiling this encyclopedia was left to the Strategic Vulnerability Branch (SVB) within the Air Force's Air Intelligence Branch. While it was renamed several times, as the Air Targets Division, the Deputy Directorate for Targets, and then in the late 1950s the Directorate for Targets, the aims were the same as during WWII. The mission was to make, as SVB's director of research James T. Lowe, put it, a "pre-analysis of the vulnerability of the U.S.S.R. [and other countries] to strategic air attack and to carry that analysis to the point where the right bombs could be put on the right targets concomitant with the decision to wage the war without any intervening time period whatsoever."[56]

Armed with new "machinic" methods to process, interpret, and model vast amounts of information, the strategic vulnerability analysts examined interdependencies between elements that they could model the effects of through a range of possible future contingencies. The SVB would work according to the now well-established doctrine of precision targeting and its methodologies for target selection. While this methodology was very similar to those developed during WWII by the COA, the EOU, and the JTG, the scale and effort that went into this were unprecedented. Not only was this effort truly global, collecting and analyzing information on every country on the planet, but it was done in peacetime.

To predetermine and select targets to be destroyed in a future war with the Soviet Union, the SVB had to centralize the military's targeting functions. This was to be done through three phases. First, data had to be produced according to standardized formats on potential targets around the world, but with a focus on the Soviet Union and its satellite states, and stored in the Bombing Encyclopedia. Second, these data had to be analyzed, emphasizing the functional relationships between targets and their vulnerabilities. Third, they had to produce the necessary materials on each of the selected targets for operational planning of atomic and nuclear war.

These novel epistemic operations were made possible through configuring existing thinking and doings with new "machine methods" of information management in the targeting assemblage. Methods that could draw together information from multiple sources, code and geolocate the nominated targets, and then automate the data management systems, which could then be used in simulations to identify key target systems. As James T. Lowe, put it, "in developing it [the BE], strategic vulnerability analysts formalized a distinctive and novel set of techniques for organizing and analyzing large amounts of data on economic and social life, and for relating the present to an uncertain future."[57]

From the mid-1950s the task of collecting and analyzing targets worldwide had become too big a task for the SVB, and the organization had expanded tremendously. This small branch had evolved into the Office of the Deputy Director for Targets, which was further divided into three divisions. The three divisions now tasked with the overall mission to "insure that the Air Force has the capability of delivering the right weapon on the right target at the right time"[58] were the Physical Vulnerability Division, the Target Materials Division, and the Target Analysis Division. All three divisions did, in their own way, sociotechnical work to prepare the US for "destroying the enemy's strength and will to wage war."[59]

The Physical Vulnerability Division was in charge of determining costs and effectiveness and developing categories of targets based on their "physical characteristics and vulnerabilities to damage by BLAST, FIRE, WATER, EARTH SHOCK, PERFORATION AND FRAGMENTATION."[60] Dossiers, or what were known as air objective folders, on each individual target in the BE were developed by the Target Materials Division. These folders contained detailed information for operational planning and guidance for bomber crews: "target data sheets, navigation charts of the Target area, approach maps, aerial photograph of the target both oblique views and vertical mosaics."[61] Each target folder further contained a large map-sized folder with a 1:25 000 photo mosaic of the designated Soviet city as the centerpiece. "All the targets within the city were delineated, described, and assigned BE numbers. The folders included analyses of all of the buildings (stone, wood, brick, etc.) in the city and the type and size of nuclear weapons that would be required to destroy it." Often aerial photography was attached to the folder, but if not, a sketch drawn from other sources, most often mass interrogation programs, was included.[62]

The Target Analysis Division oversaw the updating and maintaining of the Bombing Encyclopedia, but its main responsibility was to analyze the database for "the vulnerability of military and industrial targets of foreign countries to air attack." By analyzing "the political, sociological, psychological and general economic factors affecting the selection of air targets and targets systems," the Target Analysis Division also considered what they called "bonus targets"—the "surplus value" or ripple effects of selecting and destroying a target that warranted putting the target in a special category.[63]

By this time, it had already been established that targeting was not a job that could be done by humans alone. Since its inception in 1946, the encyclopedia had been utilizing nascent computer technologies to both store and analyze the data. As Collier and Lakoff explain, information on each target and its assigned BE number were coded and punched into an IBM card, making it possible to classify, sort, and access the massive amounts of data with greater speed and efficiency.[64] In addition, it was argued, this would eliminate human error and the inflexibility of human filing systems.

Perhaps the most important part of this machine method of information management was the ability to make "punch card runs," that is, the automated querying of already categorized, coded, and organized data. The analyst would use a machine, such as the IBM 405 electric punched card accounting machine, to "make a punch card 'run' . . . that would identify potential targets based on a range of selected criteria."[65] These runs offered analysts choices between a set of targets whose destruction would provide the most economical way of destroying the enemy, enabling the analysts to pass on objectives to the Strategic Air Command to make operational plans in case of emergency.

While Collier and Lakoff conclude that this "kind of knowledge about collective existence as a collection of vital systems vulnerable to catastrophic disruptions"[66] was new, the story told so far shows that the logics the practices of the SBV rested on were not novel but emerged from the imaginaries of the "age of system" and decades of strategic bombing and targeting development. What was new, however, was the scale of its undertakings and the means by which this was conducted. Combined, this was a massive undertaking aimed at organizing and classifying enormous amounts of information on potential targets for a future war, supposedly offering flexibility for war planners while producing a particular "closed world" view.

The "flexibility" of the Bombing Encyclopedia together with its machinic methods thus allegedly offered the targeting configurations to address a problem that had not been present during WWII—the advanced preparation for a number of future contingencies.[67] In addition, this novel configuration would help to cement targeting as a technoscientific undertaking based on the assumptions derived from systems theory that there existed a functional relationship between substrates of modern industrial systems and that these relationships contained crucial vulnerabilities, which could be effected and affected by air power.

The result was not only a new knowledge of collective existence but, importantly, the design and engineering of a particular targeting assemblage that through the logics of the BE and its punch card machines made the enemy model flexible—a flexibility that would nevertheless reinforce the imaginary of the enemy as a system, structuring how the enemy was produced and conditioning the way in which war could be imagined, assembled, planned, and waged. Containing "all possible" targets and with the built-in ability to recombine and remodel the enemy system according to specific objectives in war, the BE would effectively come to serve as a "container of possibility" for warfare,[68] restructuring the world according to military targeting.

This *tour d'horizon* has so far uncovered the emergence and cementing of military targeting as a particular martial epistemology with its imaginaries of the enemy as a system of subsystems. In line with the interest in displaying targeting end-to-end epistemic operations, I will now turn to describing in detail how the martial apparatus went about producing data, analyzing the data, and interpreting the outputs, details that will give us a deeper understanding of how targeting works in ways that continue to weigh heavily on the contemporary state of things.

The Bombing Encyclopedia in Operation

Producing the Data

In the early years of the Cold War, prior to the U2 spy plane and the space-based Earth observation systems developed under the code name Corona, there was a general lack of data on adversaries. Most of the geographic, cartographic, and infrastructural target information had come from human sources and captured German libraries of now slightly dated aerial photographs and maps covering the Soviet Union and Eastern Europe. Besides these captured documents, mass interrogation of repatriated prisoners of war, defectors, and refugees "was the chief means by which US national security agencies gathered and made targeting intelligence on the communist world before systematic aerial reconnaissance became available."[69] This mass interrogation of people, through projects such as the USAF's Project Wringer, would come to serve as a large-scale technopolitical instrument whose primary function was to support the construction and production of targets and target systems contained in the BE.

The problem was how to translate subjective human memory, inevitably partial and messy, into "objective" analytical intelligence useful for strategic bombing. It was not simply a question of collecting and sorting information that would reflect reality; it had to be reformatted and reconfigured to serve the purpose of the BE. To do so, the US designed and engineered an industrial-scale human sensor apparatus that standardized and bureaucratized the interrogation of humans so that the data production machinery could turn subjective and arcane observations into objective targets. Because of the limited amount of useful information each subject carried, volume had to be prioritized for the information to be considered reliable and valuable enough to be turned into paper-based and machine-readable material.

For the process of turning human intelligence into what Eliot Child terms "datafied memories"[70] and those datafied memories into targets, the interrogators and analysts had to further flatten out the information, decontextualize it, and strip it of any subjective narrative, alongside compiling and integrating multiple sources and accounts into one machine-readable folder. Thus, behind the numerical targets in the BE were "streams of memories, objectified and made into recombinable datapoints"[71] to produce an abstract vision of the "anonymous enemy"[72] as a system of fixed architectural or infrastructural targets that made up the sociotechnical conditions of possibility for strategic bombing.

The Army Special Documents Section (SDS) had, since the war, begun to exploit captured German documents, and to a lesser extent Russian documents captured by the Germans, on the economy of the USSR. It created something called the Industrial Card File (ICF) project, which sought to complete "carding [of] all data on all conceivable types of industrial installations, including power plants, factories, shipbuilding yards, railway repair shops etc."[73] By the end of May 1947, it was expected to have reconstructed the latest German intelligence picture of the Soviet economy. Utilizing a small group of German specialists who were working on the Soviet economy during the war, the goal of SDS was to produce reports on the overall economic potential of the USSR and on specific types of industry, a total of 35,000 cards. In the final phase, atlases would be produced showing concentrations of industries and their types and capacities.

To bring the ICF information up to date, the captured documents would be supplemented by other sources of information such as Russian newspapers and by exchanging intelligence with other branches. The ICF project took off when the function was transferred to the CIA in early 1947, and the agency was tasked with coordinating the exploitation of captured Japanese and German documents and disseminating the information in the form of ICF cards.[74]

The CIA's Foreign Industrial Register was activated in June 1947 and had three tasks: (1) maintaining a file on "foreign industrial installations, developments and resources" and their functional relationships; (2) correlating these data for the purpose of economic analysis or statistical computation; and (3) "providing economic or operational details of any particular foreign industrial installations."[75] The Industrial Register, while going beyond the efforts of the BE in trying to calculate Soviet economic output, became "deeply involved in the targeting process and provided new textual, aerial and ground photos of seventy cities."[76] By the end of 1948, the basic file consisted of some 80,000 ICF cards, classified and coded into "IBM cards so that collations for product analysis [could] be made mechanically" through "special machine searches, selections and listings."[77] It proved particularly useful for the Air Objective Folders program, as the ICF had an index volume for each country, listing all plants by location, name of installation and type, "identified by a plant number, to which the ICF card and folder are keyed," as well as concisely written "plant summaries."[78]

James T. Lowe referred to these massive amounts of data produced, indexed, and classified, all according to the purpose and logics of targeting, as a "perfect avalanche of information," arguing that the problem was not so much accessing and collecting secrets but, rather, how to manage this vast sea of data.[79]

Handling the Data

As the amounts of input increased tremendously over the years, the USAF Directorate of Targets sought to increase its capacity by developing novel mathematical models and machinic techniques for the mass handling of data. Outten Clinard makes the observation in 1959, in a then-classified study, that "although it is not impossible to solve most of these problems by manual calculations, the time requirement and cost of manual solution would be prohibitive. Some sort of machine methods have therefore become necessary."[80] He continues, "When the finished intelligence is global and encyclopedic, as in air targeting, these quantities assume massive proportions, and their management requires substantial resources in time and people or machines. Since more than storage and recall of documents or even basic intelligence information is involved in air targeting, data handling techniques have perforce developed in a complex rather than straightforward pattern."[81]

These data handling techniques, or data processing as we would call it today, Clinard describes in sterile and functional language, can be logically broken down into three distinct processes: "*document handling*, or the extraction of individual data from source materials; *data manipulation*, or the consolidation and organization of data in various arrangements; and *data integration*, or the synbook of data in application to operational problems."[82]

With regard to document handling, Clinard explains that these are mostly "library-type problems" that "lend themselves to mechanized corrective measures" such as the development of the Minicard System.[83] The biggest problem that Clinard saw was with information management and the standardization of formats and data from multiple sources. This was eventually achieved, we are told, with the Consolidated Target Intelligence File, or CTIF. The primary element of the CTIF (figure 3.1[84]) was the standard form that was filled in for each target listed in the BE. It had five parts that helped to homogenize and routinize the classification work and thereby to decrease human cognitive drifting:

i. Codes for machine processing and hand processing.

ii. Information identifying and locating the target.

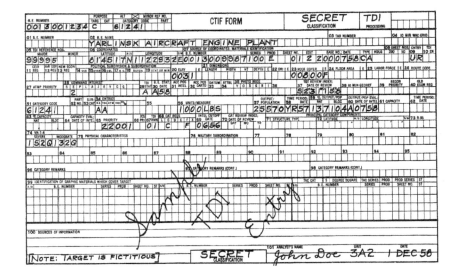

Figure 3.1

CTIF form. *Source*: Outten J. Clinard, "Developments in Air Targeting: Data Handling Techniques," *Studies in Intelligence* 3, no. 2 (1959): 100.

iii. Information on the category of the target and its individual characteristics within the category.

iv. References to graphic coverage on the target.

v. Sources.[85]

In addition, Clinard explains, the CTIFs were stored on magnetic tapes and were "susceptible of rapid and complex manipulation in electronic data-processing machines for a wide variety of purposes."[86] These included, for example, effects analysis but also, importantly for creating lists or target sets, showing "the dependence of certain plants upon the products of others and for pointing up methods of disrupting production."[87]

The CTIF marked a significant advance in data handling techniques. Providing analysts with an up-to-date file of target information, facilitating the manipulation of large amounts of data, and through its queries producing answers to complex problems at machine speed, they allegedly constituted a "giant step in facilitating mechanized support of target analysis."[88] Indeed, as Kevin McSorley argues, the CTIF and the BE provided military planners with a standardized but flexible mechanized system for planning.[89]

Although Clinard argues that the CTIF created numerous advantages, the cumbersome work of "coding much of the information and translating it into the precise language required for machine handling"[90] continued to cause difficulties. Especially, he argues, these more or less mechanical functions may cause analysts to adopt a "mechanical approach to the information," so they must be "alert not only to the facts they are recording but also their meaning" or else "be in danger of losing the feel for intelligence on which so many of their judgements must be based."[91] What Clinard is in fact arguing here is that the move toward ever more machine processing creates the danger of determining outputs, cancelling out the all-important human gut feeling and judgement that most of intelligence in the end rests on.

It was a culture clash between machines and humans, but in many ways it represented, as Packer and Reeves have argued in another context, a "new means for carrying out the immaterial labour of *perceptual, mnemonic, and epistemological labour*, the work of sensing, remembering, and knowing."[92] A configuration that was gradually stabilized through design and engineering and "dances of agency[93]," which not only altered the capacity to produce things but also created new relationships largely reorienting and determining what, how, and why things should be produced. As such, the CTIF and its data handling techniques were not only bureaucratic tools but functioned to homogenize behavior, routinizing work and reducing cognitive

The Enemy as a System: The Geopolitics of Strategic Paralysis drifting, which also served as important steps toward the martial dream of automating data processing, knowledge production, and decision-making.

The "data problem" would be further amended by the novel Air Force Intelligence Data Handling System, or System 438-L, part of the so-called Big L systems,[94] which would function as "an integrated system to accept information from any and all sources and to organize, store, manipulate, and disseminate it without the limitations of capacity and speed inherent in present practices."[95] The 438-L would further develop and integrate "man-machine techniques, procedures and equipment for increasing the effectiveness of the intelligence process."[96] In particular, the 438-L would make it feasible to develop a rich indexing system, related not only to specific documents but also to key words on individual pages, enhancing the "ability of analysts to make subtle correlations of data and develop significant interrelationships."[97] Kenneth Johnson went as far as to call the 438-L system a "quantum jump ahead" for targeting intelligence,[98] foreshadowing today's so-called big data analytics or artificial intelligence.

Intelligence analysts have long recognized that "bits and fragments of information about persons, places, things, and movements can when assembled, analysed, and synthesized enable us to make a sweeping end run around a formidable security barrier."[99] What is new, Johnson argued at the time, "is that science has come up with the technology that will permit us to use this practice on a scale and with a speed never before possible. The exploitation now made possible of the vast amount of data already on hand in different forms in many agencies offers immediate promise."[100] This required a "high degree of rapport between analyst and programmer,"[101] patiently working together translating the world into machine language and back again for human interpretation.

Operational Models

This mechanized support meant that the model of the enemy had moved from simply being an abstract representation toward an automated instrument for queries. Rather than a picture of the world, it was now a model of the world, and the key notion moving from a picture to a model is, as Hunter Heyck argues, manipulation.[102] Heyck contrasts the manipulability of scientific models to Bruno Latour's notion of *immutable mobiles*–things or objects that more or less form a stable network of associations. They are mobile, in the sense that they can be moved around, but also immutable, in the sense that they hold their shape when circulated. The map is, for Latour, the quintessential immutable mobile, but (arte)facts, texts, and also humans can be put into this category.[103] However, these objects are not instruments for the investigation of the world, but a result of it—they are pictures, rather than models. In contrast, Heyeck argues, models should be thought of as "manipulable mobile[s]: a mobile denotational device that is simultaneously a simplified representation of the world and a rule-governed instrument for inquiry."[104]

In this sense models are not only inscriptions of a particular part of the world, a simplified version, but also representations with instrumental experimental utility; they can be inspected, tested, simulated, tweaked, and reconstructed. Thus, these models are not merely "boundary objects," acting as boundaries between different societies and cultures,[105] but are performed or enacted into being,[106] functioning more as world-making engines than cameras for the world.[107] In short, they are operational.

In line with the theoretical outlook from the opening chapters, Heyck argues that manipulable mobiles are not only fluid and changeable through their own workings in a particular environment or how they can be defined differently by different users. To him, they are designed, engineered, and made to be malleable or manipulable according to specific rules, rules that make the models open to experimentation and transformations through manipulation. Models are thus not "fluid" by nature but contain a "bounded manipulability" that makes them open to manipulability governed by rules.[108] As such, models are products of human convention and mathematical, statistical, and computational logics and rules. They must conform to those rules and offer correspondence with the world they represent, making their usage as an instrumental tool possible, but at the same time they can only make sense within those rules.

Unlike Madeline Akrich's notion of "scripts," which suggests that all technical objects are inscribed by their designers with predetermined settings, or parameters for user action, which define the relationship between users and objects,[109] Heyck's bounded manipulability does not only configure the user but, also provides the user with particular rules for the manipulation of the model or object itself. So rather than binaries, between the designer and user and the technical object and user, military targeting can then be said to produce relationships between the environment, the data produced, and various modeling practices and techniques, forming particular and contextual ecologies of operations.

Transporting the notion of manipulable mobiles to this Cold War production of the enemy, we can immediately begin to see how the configuration of the BE and its human-machinic methods works as a particular framework providing the rule-governed ways through which the enemy model can emerge as an abstracted representation and, at the same time, be an instrument for epistemic operations interpreting information in context and connect it with meaning. This manipulability would be useful for nuclear war planners trying to simulate nuclear war against the Soviet Union.

Modeling and Simulating the Data

As time went on, machines and machinic calculations, modeling, and simulations came to take on a more prominent role in targeting, not only in the developments of target sets through tracing causally linked interrelationships, and to simulate the course of a potential future war, but also increasingly in the suggestion of targets. Apart from data handling techniques and the CTIF, targeting had also built "a series of mechanized analytical techniques as an aid to its intelligence production." In the world of atomic and nuclear bombs, targets were now increasingly "rated according to the immediacy of their potential threat to the United States and its allies."[110] In a future war, which was deemed to be nuclear by the late 1950s, all that mattered was the precise calculation of postattack nuclear capability. Estimations would simply not do; destruction had to be assured. Targeting would thus assume a much more active role as one needed to simulate and provide some sort of "measurement of the degree to which offensive and defensive plans can be implemented or disrupted."[111]

The atomic, and later the nuclear, bomb meant that the Soviet economic-industrial complex lost some of its importance for military targeting vis-à-vis their military forces and, in particular, their nuclear delivery capability. The enemy figure was no longer a system of industrial infrastructures but a system of nuclear infrastructures, and this brought with it changes in what was to be targeted. Although the logic and rationale of targeting stayed the same, the enemy still being imagined as a system, there was a change in emphasis toward the effects of attack and translating the physical destruction calculated into postattack operational ability. "The criteria for the selection of targets and target systems lie in the implications of these effects; and in this sense effects analysis is the mainspring and director of target selection."[112] In other words, the US needed targeting techniques and methods that could not only select targets and target sets according to strategic bombing theory but also to consider US attack capabilities for "determining the degree to which enemy operational capabilities were affected."[113]

The atomic bomb had changed the calculus of targeting, mainly because of the "greatly compressed time factor," and in 1962 "counterforce targeting," the annihilation of Soviet nuclear and military forces, had taken over from "massive retaliation" as the preferred targeting strategy. As the Top Secret Strategic Air Command Atomic Weapons Requirements Study for 1959 stated, "The SAC targeting philosophy for the Air Power Battle, recognizing this compression of time, encompasses all targets that support directly the enemy's Air Power capability." Only when the Air Power Battle had been won, the SAC document continues, "can the emphasis be shifted to the ... Systematic Destruction of the remaining SovBlock War-making potential ... and the final blows are brought to bear against his economic base and his remaining government structure, with attendant physical, sociological and psychological effects."[114] While the enemy was still imagined as a system, the importance of and value placed on its subcomponents had changed. This would, however, not in any way alleviate the horrors of nuclear targeting, as the plan included, in horrifying details, the "systemic destruction" of urban-industrial targets that specifically and explicitly aimed at the population in all major Soviet-block cities.

The time factor was a novel problem because the targets of an attack could no longer be determined only by static analysis of the effects of an assumed successful attack but had to take into account what happened during the attack phase. As Robert Adams explained, "An estimate that attack on a system would destroy all enemy nuclear storage sites, bomber bases, bombers, missile launching sites, and missiles *without an indication of the timing of attack relative to the enemy use of these resources* provides no indication of whether the enemy has delivered none or 100 percent of his nuclear weapons."[115] It had dawned on the targeteers that one could not make a strategic bombing plan without considering the enemy's actions. A simple list of targets and their interdependencies would no longer suffice. The big targeting question was now to what degree an attack could be successfully carried out within a constrained time limit. Because of the nonexistent experience of nuclear war, no best practice existed; the US martial apparatus needed ways to model and measure nuclear war and the effectiveness of a first strike. It was now turning its eyes toward operations research and another Cold War science, cybernetics.

One such methodology and technique was called the Air Battle Model, with the purpose to "provide estimates of capabilities to carry out war plans in the face of opposing offensive and defensive air operations." In other words, it was to serve as a "high-speed electronic computer simulation of the effects of air war"[116] to model and simulate the operational plans that came out of the analysis of the Bombing Encyclopedia, with the additional mix of chance and enemy war plans and resources. As Adams explains, "After the inputs are fed into it, it works through the air war in great detail, writing up its history as it goes along." The model, we are told, was "completely mechanized," "as a kind of black box," while the inputs "may be viewed as the terms of reference of a problem."[117] Creating this mechanism, "two different kinds of data are fed in for each problem to be gamed, one representing the quantities, location and status of the offensive and defensive forces of both sides, and the other roughly the strategies (intentions and plans) of both."[118] (See figure 3.2.[119])

The air battle model was thus used to make test runs to evaluate different target systems, battle plans, and strategies that, again, were fed back to analysts to determine the effects on and residual capabilities of bombed installations. Models also brought analysts "to the threshold of a precise means for determining what items of information are of critical importance, a determination which will provide new, sure guidance to collection and analysis activities."[120]

This cyclical operation, called *sensitivity analysis*, is one in which data collected from the "real" world are reconfigured and given new purpose and importance in the mathematical modeling of possible "closed world" war scenarios, whose output further outlines the need for collection and analysis, often making the analyst aware of things they had not thought of.

As a researcher at the Naval Warfare Analysis Group stated after a series of similar games played out in 1961, "The imaginary world [i.e., the game-world] is changed by induction from observations of the real world . . . and the real is changed by manipulation suggested by deduction from the imaginary."[121] As such, it was believed that the model was and should be a good way to plan and deduce policy and action. The model thus takes a role as the harbinger of truth through which the real world is further analyzed and made actionable. Johnson, doubling down on this belief, stated confidently, "The same techniques applied in simulated problems would, of course, be applied to actual hot war problems."[122]

This conclusion, however, should be approached with caution, as the researchers themselves were well aware that "war gaming will only provide an idea . . . [the] results will not be indicative of how war will go."[123] According to Adams, there are two primary reasons for this: "first, because we know that our inputs are of tenuous validity, and second, because certain result-determining conditions must be assumed and will never be explicitly analyzable."[124] Here, Adams is echoing the conclusions reached by Sharon Ghamari-Tabrizi in her study of US civilian defense analysts in the 1950s, expressing deep uncertainty and anxiety about the validity of their models and the nonexistent referent through which nuclear war planners worked.[125] Regardless of their unease with their models and simulations, they nevertheless believed that the simulation mechanism served as the best technique

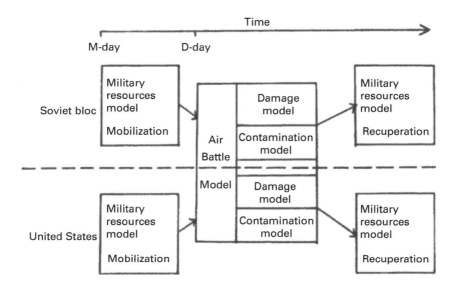

Figure 3.2

Mathematical models for Air Battle. *Source*: Kenneth T. Johnson, "Developments in Air Targeting: Progress and Future Prospects," *Studies in Intelligence* 3, no. 3 (1959): 54.

available in producing a strategy and delivering the choice of optimum target system for any strategic situation.

Humans in the Machine

Although new machine computing techniques made possible and feasible the simulation of many targeting problems, this manipulability was clearly governed by the model of the enemy produced. Thus, the way that the enemy was conceived and produced in large part determined how the simulations would and could work with the relationship between data, model, and object. But simulations did alter the divisions of labor, not pushing humans out of the loop, but introducing new spaces of war and novel "dances of agency" between human and machines bringing new forms of sociotechnical practices to the forefront of warfare. Human expertise and judgement were still vital for targeting, although the human being took on a different role in the emerging configurations of man-machine systems.

Johnson argued that humans remained the keystone of the effort, but the role of the human in the epistemic operations had changed and altered. The target analyst, he argued, "emerges as the manger, collator, and interpreter of data, instructing machines, guiding collectors, using finished intelligence produced by other analysts ... and finally producing integrated intelligence on live enemy forces on a command basis."[126] What Johnson is alluding to is that the introduction of automatic machine applications into the targeting configuration changed the role of analysts from being data processors to being data managers. Humans do not disappear in this process but reappear in different parts of the targeting operations.

As Henry Nash, a former analyst within the Air Targets Division, explained with regard to his bureaucratic and standardization work,

> In order for a nominated target to win its way into the Bombing Encyclopedia ... a Significant Summary Statement was prepared which briefly (roughly 50 words or less) described each target and its strategic importance. [We] made nominations for the integrated Air Force strategic target list and we each hoped that our targets would be chosen for a DOD strategic plan of nuclear attack. ... The complex vastness of the Defense Department prevented any intelligence analyst from determining how his work might be used by higher ranking officials. The relationship between cause and effect was obscured.[127]

Nash also explains that the bureaucracy and managerial efficiency was designed so that the analysts did not know "the larger context" of their work to produce targets, and the work to assemble the target system was reserved for higher-ranking officials. He goes on to explain that they had a so-called worst-case approach to their intelligence research, which came to take on a life of its own, "color[ing] the analyst's world-view [disposing the analysts] to concentrate on assessing enemy capabilities, primarily military, without ascertaining what enemy intentions might be."[128]

While Nash's chilling account of his work in the Air Targets Division is interested in explaining how the planning of mass killing was made possible through rational bureaucratic "distancing" and "sanitization"—in which no one is entirely in charge of decisions, so that the effects of their actions would not be fully realized—the sociotechnical systems and "dances of agency" that produced and administered targets and thus violence also had the function of reorienting and removing humans from direct killing. Although the process was still very much a human thing, the way in which humans and machines interacted in the translation processes, tuning and modifying the production and processing of data in accordance with the simulations, meant that these stabilizing dances impacted the configurations and the forms that the operationalization of the enemy could and would take—plans and actions that, as Nash explains, were far removed from the politics and intentions of the enemy.[129] It was a system that was only interested in effectivizing nuclear war, regardless of its potential horrors.

While the reconfiguration of the targeting process through machines allegedly gave analysts greater flexibility, more insights from the data, and increased efficiency of the targeting process, it also created a closed worldview in which the Bombing Encyclopedia and the models and simulations left analysts with limited options but to follow the logics of the data infrastructures and it operational outputs. Through their design and engineering, the infrastructures and practices of targeting had encoded and produced a specific kind of martial epistemology whose activation articulated and (re)produced enemy models and conditioned how wars could be imagined, assembled, and waged.

The Time Factor and the Cybernetic Other

Although it has become commonplace to argue that "nuclear war was a *tabula rasa*,"[130] and it indeed was in terms of one never having been fought, military targeting nevertheless gave atomic and nuclear war managers and planners a way to think about the unthinkable. The shift in focus and emphasis from industrial-economic infrastructure toward Soviet nuclear forces and structure did not alter the way in which the targeting process worked, only its targets. As William Walsh argued in 1948, "The atomic bomb will not necessarily create a fundamental change in the conception of target analysis . . . the difference between it and conventional weapons is one of degree which is measurable and calculable."[131]

Mathematical Monte Carlo-type models, input-output techniques such as the military resources model, systems analysis, operational war-gaming, man-machine studies, and other technoscientific innovations in modeling and simulating, became vital for the selection of targets and effect analysis of combat operations as increments in time and space became essential elements in the targeting process. The purpose of these mathematical models, as Robert Leavitt asserts, was the "selection of targets for optimum forestalling of enemy courses of action . . . assisting the intelligence analysts and planners to make better judgements."[132] While this did not alter the way the enemy was produced, it did alter the way wars were planned and the enemy operationalized. The enemy was still seen as a system-of-systems in which the most important part was their nuclear delivery capabilities, but the enemy was no longer the static, inanimate figure of the infrastructural system susceptible to attritional systems warfare. Instead, it was now a lively set of predetermined attributes with a distinct time limit attached to it. Indeed, the novel mathematical models, in particular, Monte Carlo methods, had given life to the enemy model, inserting calculated randomness and probabilities into the methods of operational research and systems analysis that had long favored prediction and clear causal claims.[133]

What had happened was that systems analysis, which formed the backbone of the Bombing Encyclopedia and grew out of the logics and theories of the enemy as a system, had been fused with another Cold War science, namely cybernetics. Cybernetics, which emerged out of the work on automated fire systems during the latter years of WWII, had taken up a prominent role in military and scientific circles during the 1950s and 1960s. However, it was largely used to think about and devise antiaircraft and antimissile systems through intricate man-machine-environment feedback systems.[134] The introduction of cybernetic ideas, such as black boxes and recurrent goal-directed feedback loops of moves and countermoves of opposing forces, into the war-gaming mechanisms of the air battle model created a hybrid of Galison's enemy figures—the "anonymous targets of air raids" and the "mechanized Enemy other"[135]—into a single system or model. Although not a fully-fledged real-time cybernetic entity, the enemy constructed through systems analysis and the air battle model amounted to a black box model that through input-output analysis and simulations created a closed system of self-regulating man-machine configurations that referred back to its targeting logic as well as the logic of the mathematical design.

In this rudimentary cybernetic network, simulations worked as a control feedback through which a human operator/manager could adjust the weights of the model, affecting both its function and mathematical logics, correcting both the performance and the rules of governing the performance. The enemy figure thus existed only in the model and only came to life through simulations, creating an interesting duality between the structured enemy as per systems analysis and the more processual enemy as per cybernetics.[136]

In practical terms, this meant that failure, disruption, surprise, friction, and chance—qualities inherent to the very practical and operational arts and acts of targeting—became core components of computing problems. In many ways, this reverberates with Wiener's claims that cybernetics was as much an epistemology for knowledge production as it was a science for the exploration and controlling of the universe.[137]

The introduction of cybernetically based simulation systems did not, however, change the logics and politics of targeting, as the mathematical models were constantly referring back to their purpose, "the selection of targets for optimum forestalling of enemy courses of action,"[138] thus furthering and strengthening the closed-world view of targeting.

Mirror Imaging

Interestingly, but perhaps not surprisingly, as the British had already used their own experience of being bombed from the air during WWII to develop techniques for targeting,[139] nuclear targeting increasingly turned inward. This had profound consequences for social and political life at home,[140] not only in the assessment of military strength and vulnerabilities as per the air battle model or more granularly in the cyborg model of human-machine systems served to us by cybernetics, but through the idea of systems vulnerability. Having developed knowledge on how to systematically destroy a modern industrial state, targeting made US war planners consider themselves through the same techniques, directing targeting and its systematic analysis of strategic vulnerabilities toward the homeland. Being increasingly seen as the only way to view the world of nuclear superpowers, targeting was in the crudest sense perceived as vital to survival. Survival, it was now argued, was dependent on minimizing potential destruction through the decentralization, distribution, and dispersal of industry and urban centers.[141]

Collier and Lakoff also make a comparison between targeting and vulnerabilities, arguing that that infrastructure came to be regarded as essential instruments or "vital systems." To minimize the potential for destruction in case of a Soviet nuclear attack, analyzing strategic vulnerabilities became a priority and led to the methodologies and techniques of "vital systems security."[142] The increasing dependence of collective life on interlinked systems, a direct consequence of the industrial age and modernization processes, Collier and Lakoff argue, led to a mutation in biopolitics toward anticipatory technologies and practices for mitigating vulnerabilities.[143] While the BE was far from a biopolitical tool in the service of population governance, with its instrumentality residing in its function as a geopolitical instrument fundamental to destroying the enemy, it clearly shows how the built world, like infrastructures, became, in historically specific ways, political.[144]

As such, the BE not only points toward the importance of archives for warfare,[145] but it shows how infrastructures have the ability to connect previous disparate practices and spaces and transform the conditions of possibility for thinking and operations.[146] The technopolitical analysis of infrastructures and the practice and process of producing targets inscribed and archived matter through a particular martial epistemology that saw the built environment as a way to activate specific war-fighting strategies, away from the immediate battlefield of the armed forces and toward the assumed webs of infrastructure that made up the enemy's system. And in the process, the BE (re)constructed operational environments and military space-times toward nonhuman elements, reinforcing the imaginary of the enemy as a system and war as a largely nonhuman affair pitting systems against systems.

Thus, targeting became a way of viewing not only the enemy and the operational environment but also oneself. While the Americans created their Bombing Encyclopedia, the Soviets had their equivalent in what John Davis and Alexander J. Kent call the Red Atlas.[147] Regardless of the extent to which the superpowers knew about each other's efforts, they had both come to the same conclusion, that a future war would be one in which targeting was crucial. Through targeting modeling and simulations, it is also plausible that that they both reached the conclusion that a nuclear war could not be won, in any sense of the word, helping to create the fragile yet somewhat stable closed world condition of mutual assured destruction (MAD).

Enemies, Networks, Effects and (Re)Construction of the Battlespace

The rest of the Cold War saw a steady, albeit contingent, growth of experimentation with different data handling techniques, computational machines, and sensors for the production and processing of targeting data and modeling and simulation of the enemy. Built through the martial epistemology of targeting and its imaginaries and configurations, Cold War innovations were all part of a fragmented process occurring within and through problems that planners observed when modeling and simulating nuclear targeting problems.[148] In many ways the logics and practices of targeting and the logics and politics of destructive functionalism showed a remarkable degree of continuation.

Although the introduction of smart bombs, or precision guided missiles (PGMs), along with electronic sensor technologies, and automated command and control systems in the late 1960s and early 1970s would make it possible to track and hit moving targets alongside fixed targets with allegedly increased precision and accuracy, the enemy was continuously modeled according to systems thinking. Greater precision did not lead to radical changes in the way that the enemy was produced, or worlds understood, but it did, in similar but distinct ways like nuclear weapons, alter the way in which the US martial apparatus would operationalize the enemy.

However, it was not until the shadow of nuclear apocalypse had faded that American target analysts, targeteers, and air force planners could again satisfy their desires and test their targeting assemblages in full force outside of computed simulations. Some four months after a "new world order" had been proclaimed by President George H.W. Bush[149], the world public was introduced to the American military's new targeting machinery—novel command arrangements, new surveillance, and offensive technologies, as well as stealth—in a highly dramatic fashion. On January 16, 1991, Operation Desert Storm was initiated against an already war-weary Iraq state, with 42 days of nonstop aerial bombardment "against strategic military, leadership, and infrastructure targets,"[150] widely destroying military and civilian infrastructure,[151] with predicable horrific effects on the civilian population.

This campaign was conducted according to US Air Force colonel John Warden's "five-ring model," which distinguished five enemy subsystems: leadership, system essentials, infrastructure, population, and armed forces.[152] Seeing the enemy as a system, Warden had updated the 1930s industrial web theory of the ACTS to reflect the 1990s strategic vulnerability of a modern enemy system. It retained the reductionism inherent in the older systems analysis—attempting to simplify complex and dynamic phenomena by reducing them down to their basic parts or functions—and the bombing campaign was set up to quickly and decisively produce a paralysis of the "central nervous system" of the enemy regime, which, it was assumed, would lead to a systemic collapse of Iraq.

With the utter destruction of Iraqi systems from above, air power enthusiasts and targeting experts had, in their minds, finally shown the power of their epistemologies. But more generally, it had given many military theorists a confirmation of the Soviet prediction of an impending "military-technical revolution" (MTR) driven predominantly by advances in information and communications technologies, enabling the seamless integration of sensors and munitions into effective "reconnaissance-strike complexes" that could strike targets at long range within minutes.[153] By the mid-1990s, advances in the precision of detection, identification, and attack capabilities also meant that "what can be seen on the modern battlefield can be hit, and what can be hit will be destroyed,"[154] giving rise to the concept of *dynamic targeting*.

More importantly, all these developments and advances in targeting organization, concepts, and technologies gave rise to a novel targeting philosophy that would have tremendous implications not just on targeting doctrine and operations but on Western warfare itself. Conceptualized by Air Force Colonel David Deptula, it become known as effects-based operations (EBO).[155]

Concerned that too much effort went into the inputs of targeting rather than the outputs—the effects—Deptula argued for changing the focus; from destruction to the aims of war.[156] Rather than focusing on massive destruction, the argument goes, forces must be able to produce a variety of functional, systemic, and psychological effects beyond the immediate physical results. Intended to be a holistic concept, it not only expanded the notion of what a target and a target system could or ought to be, but also increasingly sought to incorporate other tools of government beyond the military as means of warfare, or so-called effectors.[157] The concept ensured that everything was now a potential target and that everything could be turned into a weapon. While EBO was totalizing in terms of its target sets, and selective in terms of its application, it still relied on systems thinking in its modeling of the enemy and effects.

With the expanding notion of what a target can be and who or what can affect it, the need for new types of granular information and knowledge expanded. And with this increased dependency on information to drive modeling and operations, the problems of computing cascading effects and how to model these increased manifoldly. Thus, understanding how to achieve such effects required a different targeting configuration that was dependent on detailed and up-to-date knowledge of how various subsystems of the enemy worked and, importantly, transformed.[158] This made targeting a much more complex process that required real-time monitoring and measurement to keep up with an increasingly dynamic battlefield in which buzzwords such adaptation and flexibility were tied to other buzzwords and imaginaries such as situational awareness, information dominance, and knowledge superiority. Drawing heavily on novel sciences such as chaos, nonlinear, and complexity theories, the US military underwent a radical organizational and doctrinal transformation that rested on developing novel sensors and command, control, and communication systems and linking these with so-called effectors in the field.[159]

Doctrinized into what is now known as network-centric warfare (NCW), this represented, according to its proponents, a fundamental shift in military culture and operation toward decentralized, interconnected, and self-synchronized units operating cohesively through shared situational awareness, a transformation that, in the more optimistic view, would lift the famous Clausewitzian fog of war and reduce chance and friction.[160]

NCW formed part of a pyramid laid out in 2000 by the US military's Joint Vision 2020, whose goal was "full spectrum dominance." The enabling foundation of this pyramid, consisting of NCW, information superiority, decision superiority, and ultimately, full-spectrum dominance, was the augmented intelligence network concept of a Global Information Grid (GIG). Defined as the "globally-interconnected, end-to-end set of information capabilities, associated processes, and personnel for collecting, processing, storing, disseminating, and managing information on demand to warfighters, policy makers, and support personnel,"[161] the GIG was to make the targeting process faster, more efficient, and more effective in an increasingly dynamic targeting environment, reducing response time and increasing the operational tempo, or the battle-rhythm, to the extent that Air Force Chief of Staff General Ronald R. Fogleman, stated bluntly in 1996: "In the first quarter of the 21st century it will become possible to find, fix or track, and target anything that moves on the surface of the Earth."[162]

EBO and full-spectrum dominance, both emphasizing the ability to paralyze the enemy through overloading an adversary's perceptions and understanding of events, relied heavily on the time factor and sought to create what Brian Massumi refers to as the "force-to-own-time."[163] Harlan Ulman and James Wade, the architects behind the infamous shock and awe bombing campaign in Iraq in 2003, argue that the "target is deprivation of the senses."[164] Referring to the *rapid* in their notion of "rapid dominance," the goal is, according to them, to act within and faster than the adversary's decision-making cycle, "controlling the enemy's perception"[165] in order to make the enemy react to events that have already happened—or condition the enemy's reaction.

But as Massumi argues, this is about not just conditioning the enemy's reaction but operationalizing the future in a way that produces a future.[166] Believing that they could condition the emergent future, the US military attempted to create a battle network that would self-synchronize, rapidly and effectively distributing information across the battlespace to the smallest possible entity that could act autonomously. In other words, they imagined a network where information processing would be centralized, which in turn would enable a decentralization of battlefield operations and extend what is known as mission command, or autonomy, to units on the ground.

These readings of contemporary warfare and targeting are tightly linked to the cybernetic model of "'situational awareness"[167] and decision-making, often imagined and institutionalized through the infamous figure of the observe, orient, decide, and act, or OODA, loop.[168] While the OODA loop was developed by Colonel John R. Boyd to depict the tactical decision-making cycle of fighter pilots in the closed worlds of dogfights, the loop has recently been scaled up to accommodate larger organizational processes such as command and control and various battlefield management systems. Depictions of the loop often portray particular ecologies of operation in which the observation and orientation nodes systematically interact with the environment. However, the OODA loop itself is more tightly linked to the tactical tracking and killing of known enemy forces in particular moments in the battlespace, also known as the fighting aspect of war, than to military targeting as a martial epistemology. In other words, situational awareness and the OODA loop are less about martial epistemologies than they are about particular imaginaries of how to operationalize specific knowledges and how to network forces to enable these operations.

Nevertheless, the networking of sensors, analytics, and shooters enabled the US to create what Antoine Bousquet has referred to as a "global imperium of targeting." Aided by a semiautonomous "martial gaze," in which perception and destruction has converged, Bousquet argues, has created the possibility for a remote or prosthetic warfare that puts the whole planet into strike range within minutes. This "battlespace *in potentia*"[169] speaks directly to Frédéric Mégret's analysis of the "deconstruction of the battlefield,"[170] in which the traditional battlefield, limited in both space and time, has been supplanted by what Derek Gregory terms a multiscalar and multidimensional battlespace.[171] However, Dan Öberg argues, the battlefield is not just deconstructed but is, rather, reconfigured and reappears through various advanced analytical processes as a new abstract operational model of military space-time[172] that endlessly translates and remakes the world according to its operational logics.

Operationally, this created the need for self-adaptation through recurrent targeting cycles that would enable the networked forces to learn about the enemy and the operational environment. In practice, Nordin and Öberg argue, such a demand for continuous operations to keep up the battle-rhythm meant that there needed to be an excess of targets, leading to a situation in which the battle-rhythm "works as a 'processor' that grinds out targets and applies not so much means to ends, but the process itself as an end."[173] In this closed world model, war had become warfare, and warfare, it now seemed, had become a targeting process in all but name, creating its own self-referential and self-perpetuating cycle of operations and reconstructions of the battlespace, a topic this book will return to in chapter six.

Conclusion

While much writing on contemporary warfare has argued that it rests on the "apparatus of distinction,"[174] distinguishing between friend and foe or civilian and combatant, this chapter has shown that these apparatuses or sociotechnical systems of capture have a lengthy trajectory. However, it is not so much that targeting has been interested in particular distinctions but, rather, that it has grown out of a distinct relational understanding of the world as systems of subsystems, in which the goal has been to find the most economically effective way of destroying a modern state's ability to resist the will of the protagonist.

As this chapter has attested, the enemy-as-a-system configuration produces the world in a distinct way, reducing the infrastructures of a modern state down to their basic parts or functions, calculating their interdependencies and vulnerabilities, and from this arrangement of things creates a particular world according to targeting. Seeing the enemy as a system that can be paralyzed through the detection, classification, and destruction of vital nodes is an assumption that continues to figure the enemy. It is a world in which everything can be a target, and targets are categorized and classified according to their functional relationships into target sets, identified by a specific numerical code, located according to Universal Transverse Mercator (UTM) coordinates, and marked on special maps.

Viewing the world through infrastructures, the enemy as a system configurations fluctuate their production of the enemy according to what societies and states depend on. And it is precisely these arrangements of infrastructures that largely determine what is targeted. Each of these aspects was and continues to be part of the vast epistemic infrastructure and operations of targeting, routinizing the production of knowledge and understanding, which results in a specific and abstracted form of military legibility that is stored in the form of computer code in vast databases. These models of the enemy are not only selective representations but are also operational, functioning as tools for queries for selecting targets to be destroyed, how to most efficiently operationalize this, and what the metrics for success are, all ready to be simulated to perfect their effectiveness and be applied to the real world through operations.

As such, experiments and simulations do ontological work, providing actionable answers to what entities or parts the world is made up of and enabling an understanding of the world akin to what Steven Graham calls demodernizing by design—the simple reversal of the process of modernization.[175] And in these understandings, war is reduced to an inert object for strategists, where human costs are abstracted, sanitized, and legitimized through calculative modeling techniques.

Through its work, targeting constantly opens up new kinds of relations between the specific and the general. What is of value in this long violent trajectory of the epistemic production of enmity is always related to the object's spatiotemporal importance in the systemic or networked relation called the enemy. While the model of the enemy as a system has shown remarkable consistency over the past 100 years, what is targeted and with what has changed from industrial output and nuclear delivery mechanisms, to electricity, computers, and individuals with bombs and bytes, to finance and information. WWII saw oil refineries, military industries, and ball bearing factories targeted; the Soviet Union was produced in the same way until the atomic bomb and the time factor hampered this epistemology, but it returned in Iraq in 1991 and 2003, where electric grids were deemed to be the most vital infrastructure to obliterate.

Today, as the US war machinery has again turned its eyes and ears on "near-peer competitors," also known as China and Russia; the "enemy as a system" along with the destruction of a variety of critical infrastructures and enablers through systems warfare is back in vogue. Armed with so-called artificial intelligence, autonomous technologies, and robotic swarms, the dreams and technopolitical imaginaries of omniscience and military dominance have come back with full force, updating their command-and-control systems and operational concepts to again fight and win a war against nuclear-armed adversaries.[176]

4

Hamlets, Humans, and Worms: Computing the Environment in Vietnam

> As the battle field was not linear, there were no front lines facing each other; no enemy fixed positions; no headquarters, artillery positions, supply dumps; none, in fact, of the targets found on a conventional battle field.[1]
> — Gordon L. Rottman, United States Army, Special Forces

Gordon Rottman's perceptive comment on the problems facing the US military in Vietnam only became apparent for the Americans as the war dragged on. As a war without a front, and with limited conventional targets available, the targeting assemblage and the Bombing Encyclopedia configured through WWII and reconfigured for a potential nuclear confrontation with the Soviet Union offered few options to the US martial apparatus. Designed to provide planners and targeteers with a blueprint for system collapse through infrastructure and industrial targeting, military targeting was not configured to fight a limited war against a supposedly "faceless"[2] and amorphous enemy that offered few of the same fixed sites for attack, or as a coercive instrument aimed at changing the strategic calculus, the will, of the North Vietnamese government. In addition, military strategists could no longer rely on standard operating procedures or measure progress in terms of land seized or enemies killed, and the old "order of battle" (OB)[3] assessments did not provide military planners with knowledge of an enemy that could neither be found nor understood in the traditional way.

New ways of finding, tracking, and understanding the unknown knowns and known unknowns had to be devised. While the Cold War superpower rivalry, as we have seen, was largely was "fought" through simulations, war games, and computer modeling, Vietnam afforded a very different form of on-the-ground experimental site for targeting. Armed with the computer and a series of managerial approaches to war, the US martial assemblage turned Vietnam into an experimental laboratory in which a number of different programs and systems were configured, tested, and refined—with long-lasting devastating effects on Vietnam and its population.

This chapter will focus on three of these: the Hamlet Evaluation System, a computerized counterinsurgency reporting and measurement system to produce knowledge about the rural population in Vietnam; the Igloo White system, a command and control system aimed at computerizing and automating the battlefield with remote sensors, computers, and missiles; and the infamous Phung Hoang/Phoenix program, an effort to identify and "neutralize" the Viet Cong infrastructure (VCI) through continuous targeting cycles and kill/capture teams. While these configurations take us from the air down to the ground,[4] the chapter's narrative is not so much about these different spatial cartographies as it is about moving from systems to networks as the main martial structuring frame for the epistemic operations to make up the enemy.

What ties these systems and programs together is their various efforts to compute the environment in Vietnam and the centrality of information processing to the production and understanding of the battlespace and the enemy. As such, they all functioned as experimental sites for the translation of the environment into actionable intelligence, supporting the vision of war as something that could be controlled through managerial and technoscientific logics and rationalities.

The way computers were designed and engineered into these systems helped create and sustain this imaginary in several ways. They allowed for the practical construction of military information and control systems, which upheld the view of war, the operational environment, and the enemy as a systems or networks that were calculable and amendable to technological management. These systems were not merely human prosthetics or tools but in more substantial ways helped create, structure, and alter "reality" and the possibilities within this reality, reducing the problem of war to an issue of sensors, analytics, and outputs. While cybernetics provided feedback loops to guide operations, system analysis formalized and linked the choices about strategy to technoscientific practices by asking how a mission or problem could be accomplished with actual or potential technologies. Seeking to displace the Clausewitzian fog and friction with computer-enabled systems that could tackle complex problems, these ecologies of operations not only produced different models of the enemy through novel epistemic practices and infrastructures but also, at the same time, introduced new forms of fog and friction into the war.

While the so-called managerial approach to war in Vietnam has already received sustained criticism,[5] this chapter will show that it was not only the managerial logics that shaped the warfare but the ways in which long-standing imaginaries of the enemy and war as quantifiable entities were configured into and operationalized through different ecologies of operations. More than the managerial approach, human-machine configurations mediated, shaped, and structured the course of the military engagement in Vietnam in many ways. Moreover, the chapter will show how the epistemic and infrastructural legacies and practices of these systems continue to have a significant impact on the present in how the enemy is understood and the increasingly totalizing view of military targeting.

War Managerialism, Data, and Computers

Famously, US Secretary of Defense Robert McNamara arrived in office with a reputation as a great business manager, having turned around the fortunes of the Ford Motor Company after WWII. Armed with the latest science and the digital computer, which both drew on and furthered a view of the world as quantifiable in terms of logical and rational systems of control and communication, McNamara transformed the Department of Defense (DoD) and the way Americans approached warfare. "McNamara's early applications of computers to war were ground-breaking," Donald Fisher Harrison argues. "Using computers as an analytical tool, he soon made fundamental changes in the department's reporting techniques, as well as in the use of computer-generated data for decision making."[6]

To cement this approach to war, McNamara created a new position at the DoD, assistant secretary of systems analysis, which called for the broad use of statistical evidence as an essential tool for operational research and decision-making, all aided by computer analysis. What McNamara would instigate, and personify, is now known as the managerial approach to waging war, relying heavily on quantification, statistical analysis and models, computers, archives, and databases in an attempt to rationally manage and control war.[7] The Vietnam War was, in many ways, the first "technowar"[8] conducted through an ideology of managerialism and driven by "quantitative business-analysis techniques."[10] It was the first time computer technologies were fully integrated into nearly every aspect of the US war-fighting machine, making it a key configurational element in the transformation of warfare.

While these imaginaries and techniques have received much criticism, these approaches to war were, as we saw in the previous chapter, hardly new. Although systematization and quantification of war follow a long trajectory of military thinking, the Vietnam War does mark an important transformation in military thinking and practice.

Computers had, by the mid-1960s, expanded from being labor-saving calculators to becoming expansive multipurpose data processors generating actionable information and choices from vast quantities of inputted data. With the development of the computerized Semi-Automatic Ground

Environment (SAGE) system came innovations in digital memory formats, operating systems, and screen interfaces that rendered data streams legible to the system operators, changing both the nature of and the availability of information for war managers.[11] They became the integral infrastructure that allowed for novel ways of operationalizing the managerial approach to war. In particular, as we also saw in the last chapter, they enabled a closer integration of two sciences and ways of thinking: cybernetics and systems analysis.

Acting as information processors, computers would serve a twin purpose in the Vietnam War: first, by continuing the work done by the WWII targeting assemblage and the Bombing Encyclopedia of seeing the world and the enemy as a system of systems, producing, storing, and analyzing data from a variety of practices into knowledge about the structure of the enemy and the operating environment. Second, expanding cybernetics' teleologically oriented theories of recursive information feedback loops[12] from fire-control systems into the whole war-fighting machinery allowed for closer integration of humans, technologies, and the environment and novel centralizations of command and control.[13] This integration of systems analysis with cybernetics meant that the military would increasingly conduct operations according to rapid information feedback loops based on data generated through operations rather than the standard method of seizing and holding territory from enemy military forces. Based on a technocratic worldview and logic, configured in and through IBM microprocessors, the Vietnam War created a series of innovative and experimental computational efforts, projects, programs, and initiatives aimed at processing information to produce, operationalize, and destroy the enemy.

The Hamlets

In 1967, the US launched an automated quantification system known as the Hamlet Evaluation System (HES). The HES was developed to produce reliable information on the rural population of South Vietnam, where large numbers of the population were believed to support the communist North and the Viet Cong or at least to be opposed to the South Vietnamese Government (GVN). This was not a farfetched conclusion given the draconian measures introduced by the GVN, such as the forced resettlement of rural populations under the Strategic Hamlet Program and US "search and destroy" missions and indiscriminate bombing strategies, which had displaced and made homeless well over three million South Vietnamese. Regardless of the past, HES was built under the assumption and logic that if the GVN and the US could provide security and development (loosely defined), the rural population would accept GVN rule, and they would win the war. To deliver security and development, they believed they needed a system that could help them translate village life into knowledge and improve understanding of the dynamics and trends of the population.

Knowledge about the population was absent in large part due to the emphasis placed by the US military intelligence apparatus on "order of battle assessments" that focused on the combat effectiveness of enemy military forces (the North Vietnamese Army) and the famous "counting-regime," which was mostly interested in dead bodies, sorties flown, and bombs dropped. The little intelligence that existed on the rural population was either of Vietnamese origin, which the Americans did not fully trust, or in long narrative (subjective) form, mostly produced by CIA operatives in the field. The latter was viewed by the US military as of little value to commanders on the ground, lacking actionable intelligence and not providing a statistical basis for analyzing trends and dynamics over time.[14]

The HES was to amend this through the creation of a computerized system, designed and engineered to systematize and regularize human observations made at the hamlet level. It was to track a broad range of so-called security and development factors, from enemy activity and friendly presence to food, health, and education. The HES was to be an intelligence source whose promise rested on its ability to store information and its novel capacity to generate monthly statistical reports and high-resolution computer-generated maps featuring every hamlet in South Vietnam. Through its comprehensive statistical output, the hope was that this would make the South Vietnamese society "legible."[15]

It was a system to identify and sort, count, and classify the population in order to provide a statistically "objective" picture of the situation. Although not the only data collection system in play in Vietnam to systematize the conflict, HES nevertheless stood out as the most comprehensive, wide-ranging, and ambitious program. In short, the HES, in Oliver Belcher's words, "enacted a new way of understanding the Vietnam population."[16] While its immediate goal was to make the society comprehensible, it was to serve as an important measure of effectiveness in determining whether the US pacification strategy was succeeding or failing. In addition to providing intelligence on the pacification effort, the HES allegedly allowed military commanders to better see less secure areas and plan and conduct their security-development efforts and military operations accordingly. The HES was to provide an operational model of the Vietnamese population.

Designing the HES

The HES was part of the broader and long-running so-called "pacification," or counterinsurgency, effort in Vietnam. Pacification's goal was to secure and protect the rural population from insurgents and meet the needs of the people in order to, the logic went, deprive the insurgents of their popular base and generate rural support for the Saigon regime, which subsequently would lead to victory. Thus, the HES was designed and engineered from the logics and rationalities of a security-development nexus with little consideration for the grievances toward the regime in the south or the general hostility toward the American forces.

The system was developed and became operational in January 1967 under the administration of the newly created civil-military advisory organization called Civil Operations and Revolutionary Development Support (CORDS). According to its head, Robert William ("Blowtorch Bob") Komer, the HES "was designed specifically to overcome the flaws inherent in previous more subjective efforts to assess what was really happening in the countryside—largely narrative reports based on Vietnamese sources that had proved consistently overoptimistic,"[17] a statement consistent with the managerial approach, trusting computers and "objective" statistics more than seasoned CIA operatives in the field and their "subjective" views.

To overcome the subjective reporting and systematically collect relevant pacification data that could be churned into objective statistical reports for operational management purposes, Komer argued, "stress had to be laid on relatively simple quantitative techniques . . . the systems had to be designed realistically for input by relatively unskilled field operatives . . . and then not to permit them to be changed as they travelled up the line."[18]

The result of this was an automated file structure called the HAMLA, into which the HES reports from the field were entered. Subsequently, automated programs applied rating criteria from experts uniformly, converting or translating the advisors' judgements into a score using a standard statistical analytical technique. Although the HES still had to rely on "relatively unskilled" humans in the field, the turn from human-based to computer-based analytics would make this endeavor objective, it was argued. As such, the computer was seen as the instrument that turned subjective human input into objective output. So important was the electronic computer to this system that, according to Oliver Belcher, it "was the apparatus that made this population-based system possible."[19]

Indeed, the HES differs in significant ways from its predecessors, not only in scale, effort, and centralized bureaucratic management under CORDS but in the way it was designed and implemented through the use of computers to distill large quantities of information in order to identify "problem areas for management purposes."[20] It was thus more than a progress reporting system, its chief purpose being to centrally control and manage the nascent, vast, violent, and messy experiment called pacification. The HES unified the reporting and tracking of pacification by creating a purpose, methodology, and human-machine infrastructure leveraging the emerging computer technology of the time that purportedly allowed for accelerated and expanded analysis of information down to the hamlet or village level.

Although the computer is a central and essential element in the story of the HES, the configuration of human-machine-environment is the key element in the story told here. The HES was configured in such a way that humans served as "environmental" *sensors* while computers carried out the *analysis*, whose *output* was in turn interpreted by humans in order to act in and on the "environment." The end-to-end operations to generate the HES model rating are therefore an interesting example of configured divisions of labor between humans and machines and the resultant decrease in cognitive wandering.

Human Sensors, Computer Analytics, and Illusions of Objective Outputs

The role of environmental sensors was trusted to the newly created district senior advisors (DSAs). Each advisor was responsible for several hamlets and provided monthly reports on each of them. Using standard-format multiple-choice questionnaires, known as Hamlet Evaluation Worksheets (HEWs), or later, the HES Question-Set,[21] the advisors coded and scored everything from location, based on Universal Transverse Mercator (UTM) coordinates,[22] to type of hamlet into a matrix (figure 4.1[23]). This matrix consisted of eighteen specific security and development indicators—enemy presence, economy, health, education, and so on[24]—according to a simplified scoring system involving five alphabetical predetermined model scores (A–E) that reflected the numeric responses to questions posed by the advisors. For instance, is there a school in this hamlet? Do more than 90 percent of the children attend? Yes = grade A. These rigid quantitative criteria selected from predetermined multiple-choice ratings in order to arrive at an overall score would, according to Komer, not only be objective but lead to comparability of hamlets as well as being able to discern dynamics over time in order to predict where to focus their military effort.[25]

By creating a baseline or normal of the situation at the hamlet level, it was believed, progress or decline could be visualized over time. In addition, the use of standardized multiple-choice forms and automated data

Figure 4.1

Hamlet Evaluation Worksheet sample. *Source*:
Thomas L. Ahern, *CIA and Rural Pacification in
South Vietnam (U)* (Langley, VA: Center for the
Study of Intelligence, Central Intelligence Agency
(2001): 419.

processing minimized human workload and facilitated storage (on magnetic tape), tabulation, and analysis of reported information. Although the HEW contained a remarks section, where the DSAs could record additional or explanatory information or raise questions about the process or any particular hamlet's evaluation criteria, little is known about to what extent these were taken into consideration before or after the scores had been fed into the automated processing system and the analysis had been done. However, given Komer's stress on objectivity and comparability, it is unlikely that any subjective remarks from advisors in the field would be taken into consideration regarding the final outputs from the automated processing of the files, which on average was generating 90,000 pages of reports monthly.[26]

The HEW was in many ways similar to the CTIF files of the BE, not only being a bureaucratic tool, but homogenizing the behavior of the human sensors, routinizing their work, and decreasing the cognitive drifting and subjective interpretations of individuals. In the original 1967–1970 version of HES, it was not until the human sensor scores were fed into a "memory bank"—which did automated data processing—that the computer did actual work. However, the computer played a significant role in the production of data as it conditioned the ways in which the HEW was constructed and how sensors went about their reporting.

Once data had been produced, the automated data processing analyzed and compared data month by month as well as the functional categories of security and development. The outputs were the Hamlet Evaluation Summary Form (HESF), a computer-generated form containing identification data for all hamlets in the district, used to aggregate and report evaluations, and the Hamlet Classification Form (HCF), used to record and report the security category and the revolutionary development classification.

Several independent assessments and reviews, as well as general critique of the system, led to reconfigurations of the HES, first in 1970 and then again in 1971. The "independent" HES Study report from May 1968, sponsored by the Army Concept Team in Vietnam, reported on the difficulties with "attempts to translate events that occur in the hamlets into the terms of indicators one through eighteen," arguing that the events and indicators had to be interpreted by the subjective human sensor. "The method of interpretation is the creation of the individual advisor," and the quality of this interpretation depended on "his intelligence, education and experience."[27] The report thus pointed out what was obvious to the critics of the HES, but concluded that "the HES is currently a very useful tool for measuring the overall progress of pacification in Vietnam. Its potential—as a hamlet information bank and subsequent use as a management device—is even larger."

Regardless of the input-analysis-output process problems, it was "non-use for management purposes"[28] that seemed to preoccupy the study. The reasons for this nonuse, the study found, fell into five rough categories:

First, the information in HES was too voluminous to extract and rearrange in forms necessary for management use. The maps furnished by MACV were generally said to be inadequate for such purposes because of their small size. Second, and related to the first, corps personnel claimed that the rating scheme was too complex of the higher ranking for management use at corps. Third, few of the higher ranking officers at this level had familiarized themselves with the types of information contained in HES. Fourth, there was a noticeable prejudice against a machine record report. Machine records could not tell you what you need to know and that the type of war Vietnam was not amendable to "computerization." Fifth, the ratings themselves could not be trusted according to some corps interviewees.[29]

In order for the HES to be used as a management device for the pacification effort, the study recommended a number of revisions, from turning the reports into an easily and quickly understandable format to be "used and mitigate biases against it," to instructing personnel on how to utilize the output, to how to present breakouts in graph form, to making the overlays larger and suitable for use on "pictomaps."[30] This would increase the utility of the HES as a management tool, a measuring device for cost-effectiveness analysis, identify priority areas for military operations, and measure the overall pacification effort, the report concluded. The recommendations from the 1968 HES study were thus mainly about how to make the objective output from the system better for management purposes and not to challenge the subjective input from advisors, the subjective-objective premise, or the overall validity of the system itself. The authors seem to conclude that quantification is objective regardless of the input, thus catering to the continuation of "quantitative business-analysis techniques" being applied to the war.

The HES/70 was nevertheless reconfigured on the basis that there was a need to involve a more "detailed and objective" question set and scope for responses,[31] expand the database to include more functional areas of pacification, increase the uniformity of the data sets, and increase the utility of the system by designing reports specifically for commanders in the field. It was believed that more data would allow for more complexity to be appreciated and increase the utility of the systems. The HES/70 introduced analytical models to reflect the perceived situation in Vietnam—which at that time was more concerned with the political activity of the VCI than the military operations of the North Vietnamese Army (NVA) or the Viet

Cong—expanding the security-development matrix into three macro-models, security, socioeconomic, and political, further divided into models, which were again divided into submodels linked to component questions.

To make the HES less subjective, one major change was introduced into the system. The advisors were stripped of the job of undertaking the hamlet scoring themselves, leaving them simply answering the predefined questionnaire. The scoring of indicators, A to E, was now to be "done centrally by a mathematical weighting formula,"[32] an algorithm built upon traditional Bayesian statistical inference. The result of this reconfiguration in the design and division of labor between humans and machines was that the HES/70 and the HES/71 produced different absolute assessments due to the different metric inputs and statistical processes employed to organize and report them. Trend lines, or the outputs over time, were nevertheless compatible, although there are reasons to believe, as former CIA operative Thomas C. Thayer notes in his *A Systems Analysis View of the Vietnam War*, that the 40 percent increase in pacified hamlets "may come from accounting-type changes not actual pacification improvements."[33]

The HES was not the only data reporting system designed by CORDS, but it was the most comprehensive one and provided "much of the basic data for other systems."[34] CORDS created over a dozen others and fed their data along with those from the HES into a comprehensive Pacification Evaluation System (PACES) that could be combined with other information streams and systems used by U.S. Military Assistance Command Vietnam. All the data from these diverse systems was fed into the National Military Command System (NMCS) Information Processing System 360 Formatted File System, commonly referred to as NIPS or NIPS 360 FFS. It was an advanced data management system developed in the 1960s under contract with IBM and operated on IBM System/360 and System/370 computers.

As the US National Archives writes in its introduction to the electronic records relating to the Vietnam War, the system provided powerful, efficient, and flexible data management support to a wide variety of users with the ability to "structure files, generate and maintain files, revise and update files and data, select and retrieve data, and generate reports" in simple or complex arrays on a variety of output devices.[35] Through NIPS, analysts were able to generate a number of monthly products that could track hamlet status, indicator changes and correlations, trends, and so on in both textual and graphic formats. However, this would only serve the purpose to feed the belief in quantification, not to portray the hamlets in a more genuine way.

The graphic format that was used was the so-called pictomap—a topographic map in which the standard mosaic was overlaid and converted into interpretable colors and symbols that showed the status of each hamlet throughout South Vietnam according to its modeled reality. These maps were, in fact, early versions of geographic information systems (GIS), designed to create, manage, analyze, and map various forms of data, simplifying the problem space through several prefiltered observations (surveys) matched to a planned problem-solving set (in/security) with predefined logical map layers to identify a coherent spatiotemporal narrative. While these were intended to provide a powerful visual tool for war managers and commanders by showing progress or areas in need of attention, the spatiotemporal thresholds they portrayed suffered from these orderly but fixed-in-time means of relating data about the environment to an overall model.[36]

There is no doubt that the HES, in its ability to show progress modeled against a sociotechnically produced "normal," worked as a legitimizing, rationalizing, and meaning-making tool for continued US engagement in Vietnam. It did, however, create a "reverse quagmire" or "mission creep" situation in which the HES, with all its faults, created a "veil of objectivity" of progress at the rural level, thus giving the impetus to continue on a "winning" path. This illusion of victory dragged the US even deeper into the war with devastating consequences.

One of the problems was that the HES system represented a snapshot of security and development at one point in time: when the advisor was present and carried out the questionnaire. Thus, the aggregated monthly statistics and trends over time only provided a series of snapshots, aggregated into a baseline or normal state of affairs, that created an "impression of steady progress and widespread GVN influence among the people of South Vietnam."[37] The statistics created an illusion that freedom from attack or harassment equaled an indicator score of A, or pacified, while it in many instances would hide covert VC influences and exploitation, a situation in which the modeled reality deceived the US military. What was created was not an objective reality but a space in which the particular was allowed to stand for the aggregated in a way that was at the same time convincing, but also meaningless.

The HES influenced and shaped military strategies and operational responses to perceived trends and dynamics, such as continued search and destroy missions and the infamous Phong Hoang/Phoenix program. Thomas Ahem argues that although the original subjective HES/67 "had some potential for qualitative measurement, CORDS management increasingly applied it in the sterile, quantitative way used to evaluate military operations [and] with its conflation of the terms security and control, in effect abdicated the whole question of peasant loyalties."[38] During the course of the war, the HES had moved on from its origin as an epistemic instrument for subjective interpretation to becoming a way of presenting a statistical reality model that would not only (mis)represent reality but also drive the US war effort in specific destructive ways.

Although there was at the time a lot of criticism of the HES, it lived on because it simply was the best system the US military had according to their management logics and system-oriented perspectives on conflict, where anything from logistics to the political allegiance of rural populations, it was believed, could be quantified, managed, and controlled. Indeed, the man in charge of the HES was fully aware of its limitations but nevertheless continued to believe in it. Komer, in a conversation after he had stepped down as the head of CORDS, stated, "HES was full of weaknesses. We wish we had better measurement techniques. But given the situation at the time could we have done better with the resources at hand? I don't think so. Moreover, has anybody come up with anything anywhere near as good in the entire history of the Vietnam War?"[39]

The main problem, as Komer saw it, was not that the statistical outputs were wrong, as such a large body of data would statistically flatten out underreporting or overreporting, but rather that the "'gold mine' of raw data on various facets of pacification impact ... has [not] yet been analyzed at depth."[40] Lamenting the lack of resources and capability to analyze all the raw data collected, Komer's remarks are strikingly similar to the problem that surfaced within the US military in the 2010s, when big data collection had largely become an end unto itself and engendered the general problem of "swimming in sensors, drowning in data."[41] The problem Komer failed to see, on the other hand, was that the US martial machinery's assumptions about what can be quantified and the subsequent design and engineering of humans, technologies, and the environment led to a system that could not offer anything other than what it did.

In 1970 Colonel Erwin R. Brigham, in charge of the technical aspects of the HES, wrote an article laying out the HES's strengths and to some extent its weaknesses. Although Komer would argue that the HES was objective, Brigham openly asserts that "basic data depends on the subjective evaluation of district advisors about problems that are difficult to quantify,"[42] adding a caution that "it must be used in full knowledge that it is a pioneering effort and that its data is imprecise."[43] Brigham, aware of HES's limitations and problems, nevertheless argued that "this system was designed to answer questions that can be answered in no other ways" and that it "is proving to be a reasonably reliable technique for measuring *those key aspects of pacification that are measurable*—hamlet population, security and development."[44] He ends his argument by stating that it would be impossible to say anything about Vietnam without an instrument such as the HES, "which develop[ed] a national mosaic composed of individual hamlets."[45]

It is perhaps these words from Brigham that best sum up the problem with the HES as well as the general problem with quantifying war. One ends up measuring that which can be measured, while leaving out that which is not as readily measurable, quantifiable, and computable. Those things that cannot be measured, quantified, and computerized, such as the intentions and will of the adversary or the loyalties of the rural population, are often the things that are of great importance but that get neglected. Much, it seems, gets lost in translation. But at the end point of these translations are tremendous amounts of violence, death, and despair.

The Model as Reality

The HES was not the only tool used, and it evolved over time alongside many other pacification tools, forming part of a larger assemblage of diverse practices and infrastructures that fed into each other. This assemblage was aimed at grasping and making legible the enemy, the battlefield, and the environment the Americans found themselves in, but instead created an illusion of the reality in Vietnam as something that could be made into a manageable object open to technical solutions. Despite doubts about the subjective inputs, the HES managed to show seasoned commanders that computer-generated outputs could be objective and capable of making the battlefield "transparent," to the extent, as Belcher argues, that the subjective judgement of experienced commanders with battlefield experience was replaced in favor of a computationally based view within the US military at all levels,[46] a process similar to when civilian scientists took over nuclear war planning from the generals.

Rather than a standard process of simplification to make things legible according to the logics of modern states,[47] the HES was a system that was designed not simply to reduce complexity but allow for it. As Oliver Belcher argues, "The aim was not simplification, but rather allowance for the *complexity* of population dynamics to be appreciated on its own terms in 'real-time', calculated through Bayesian statistical analysis."[48] While one can question Belcher's usage of the term "real time," as the HES mainly produced monthly reports from snapshots in time, the HES created what Belcher calls a distinctive "view of below," allowing system analysts within the US military to "disclose dynamics, trends, and even make data-based predictions in the pacification war."[49]

This view of below, Belcher contends, should be contrasted to the specific "view from above," which we encountered in the previous chapter. The HES offered a whole new way of seeing conflict, one that was distinct from the abstracted aerial view of battlefields and fixed industrial sites, emphasizing the computationally processed combination of hundreds of subjective observations from human sensors decontextualizing and translating rural life into machine-readable format, turning the population and the environment into an object for manipulation. While allowing for complexity, it was a structured view of temporal complexity, translating a selective aspect of reality and producing a computerized model from inherently subjective observations.

From a technical viewpoint, the problem, as algorithmic modeling expert Leo Breiman argues in a critique of traditional statistical modeling, is that traditional models are fitted to the data to draw quantitative conclusions. The conclusions are then "about the model's mechanisms, and not about nature's mechanisms."[50] Although Breiman's critique makes a case for his own algorithmic modeling, his argument is particularly salient when considering the HES and the belief in it. "The belief in the infallibility of data models was almost religious. It is a strange phenomenon—once a model is made, then it becomes truth and the conclusions from it are infallible."[51] By often mistaking or confusing the modeled mosaic for a complex reality, the HES nevertheless produced a set of actionable choices measured against a previous normal that made the US military able to act.

However, this view of below, while allowing for complexity, is also, as Donna Haraway has theorized, a view "from nowhere" serving as "the god trick," rendering all other positions invalid and subjective, and it materializes that which it embraces—or in this case makes the intangible population into a technical object of knowledge and action.[52] What matters is the things that are calculable and computable, that which is computable is also actionable, and so the abstractions produced through the translation practices of the HES are the things that are operational, linked to desired managerial effects that it was believed could not only be identified through a series of steps but also calculated, computed, modeled, and simulated with precision. The HES was therefore a key part of how "war is sustained through knowledge systems and frameworks of understanding ... that provide the conditions of possibility for war as a transformative vehicle."[53]

In the effort to create a war-winning system of calculation, the HES was close to what Edwards has referred to, in another context, as a "closed world"—"an inescapable self-referential space where every thought, word and action is ultimately directed back toward a central struggle."[54] Echoing Edwards's formulation, Knorr Cetina argues that machineries of knowledge construction also often become "self-referential systems that orient more to internal and previous systems than to the outside environment."[55] As this exploration has shown, this is in many ways also the case of the HES, catering to the functionality of the system referring back to its imaginaries and logics. The HES is an empirical example of how different human-machine-environment configurations structured the bureaucracy of war, how computers were imagined as epistemological tools that can translate the world into objective truths,[56] and how this actionable "view from nowhere" shaped the dynamics of conflict.

Although never realized to the full extent of what we see today, the HES serves as an important precursor of today's use of big data, advanced analytics, and experimental computing for targeting, offering a selective view of the complexities of sociopolitical life. In large part, the HES follows the larger trend from the Bombing Encyclopedia of viewing enmity not as a narrative subjection of the Other but as the creation of an abstract model and operational environment, digitally stored on mainframes or magnetic tapes. While the BE used computers to disclose functional relationships and interdependence among filed targets, the HES was supposedly capable of disclosing biopolitical relationships between seemingly disparate phenomena, creating a pattern-of-life and system-of-systems understanding at the level of hamlets and populations. Its legacy lives on. Both the logics and rationalities its rests on but also, more importantly, its design, its data infrastructure, and its epistemological dreams.

The Humans

In late December 1967, the government of South Vietnam announced a reorganization of its war effort against the Viet Cong insurgency, grouping all South Vietnamese counterinsurgency activities under a new program known as Phuong Hoang. In response to the South Vietnamese move, American officials in Vietnam began referring to their own counterinsurgency coordination efforts by the name that they deemed the closest Western analogue to the Vietnamese mythical creature, the Phoenix. The infamous Phoenix program, designed by the CIA, in close cooperation with the GVN, was a covert intelligence operation, engineered to identify and neutralize the VCI through killing, capture, kidnapping, assassinations, torture, so-called induced surrender, or simply disruptive action.

To defeat this elusive infrastructure and insurgency, the US martial apparatus deemed it necessary to come up with a new way of envisioning and producing the enemy than those the intelligence apparatus had been drilled in since the 1940s—the enemy-as-a-system targeting configuration and the concept of order of battle used to judge the combat effectiveness of enemy forces, dividing its military forces into different interrelated elements or factors, including their composition, disposition, strength, and so on.[57] While OB was used by the Combined Intelligence Center Vietnam (CICV), the problem was that the VCI and its militant arm did not provide any of the usual elements for the US to conduct OB assessments on due to a lack of visual strength and reliance on covert operations.

A secret CIA intelligence memorandum from May 23, 1967, calls the VCI an "intricate network" of political operatives in the rural areas, to which the population "provides both the local manpower and resources, without which, the insurgency would collapse."[58] The success of the insurgency, the memorandum asserts, was therefore directly connected to the performance of these operatives or cadres. These, it was argued, were the insurgency's "greatest source of strength" but at the same time also their "greatest area of vulnerability, for losses among this group by death, capture, or defection constitute blows at the vitals of the entire movement,"[59] A logic remarkably similar to that of the enemy as a system.

The problem with delivering blows to the vitals of the VCI was, however, that the communists, according to the memorandum, were "highly successful in enshrouding these cadre from our eyes." Unmasking these cadres would "require a) the collection of precise, timely intelligence on the targets, b) the ability to collate and process rapidly the exhaustive data that we do acquire and, c) the means to take prompt, direct action commensurate with the identified target."[60]

Assembling the Phoenix

Although the Phoenix program was initiated by CORDS in December 1967, it drew on preceding efforts by both the GVN and CIA. The GVN had established several programs to suppress domestic opposition, and the CIA had recruited counterterror (CT) teams in the late 1950s and 1960s under what was known as the Mountain Scout Program. These were teams consisting of small numbers of men trained to conduct covert assassin missions into enemy-controlled territory. When intelligence on the identities and locations of enemy operatives was received, a CT team was dispatched to kill/capture them, with the hope that these missions would wear down and destroy the VCI.

In order to understand and gain so-called actionable intelligence on the network of communist cadres and agents living and operating undercover in rural areas, five programs had been initiated prior to 1967.[61] These included the Hamlet Informant Program, which recruited informants throughout South Vietnam, Agent Penetration Operations, aimed at penetrating the VC organization, and the CT teams rebranded into so-called provincial reconnaissance units (PRUs), which conducted special operations against the VCI to collect intelligence and kill/capture individuals. The two other programs, the Census-Grievance program (C-G) and the province interrogation centers (PICs), were crucial to the gathering and compiling of information to unravel the clandestine VCI and establish a continuous targeting cycle that was hoped would disrupt, dismantle, and destroy the growing insurgency in South Vietnam.

The C-G program was a Vietnamese initiative established in June 1964. In a memorandum to Secretary of Defense McNamara, George Carver, special assistant to the head of the CIA for Vietnamese affairs, explained that the C-G had two purposes: "the main *overt* purposes of which are to assist Province Chiefs in determining the political sympathies of the province population and to establish a mechanism for the articulation of aspirations and redress of grievances. A *covert* purpose and an important product of the program is to develop information from hamlet residents and villagers on the local VC organization and activities."[62]

The C-G program worked by dispatching personnel to villages and hamlets under government control. After taking a census of the population, the C-G team would conduct daily compulsory interviews with every adult inhabitant, compiling detailed information on every resident, including data on kinship, political and religious affiliation, and property ownership. The interviews were, much like the HES, conducted using a series of standard questions so that they would, in theory, be consistent. A person's answer could then be measured and compared against others. This analyzed information would then often be used to pressure families and communities to comply with government directives and ferret out information about the VCI.

These detailed, but also highly subjective and partial, records of the population were crucial for supplying information on the VCI and the development of the Phoenix program. Color-coded maps were also made to illustrate and visualize each hamlet's loyalty to the South Vietnamese Government, as well as names of families who were VCI members or sympathizers. Under the guise of addressing grievances, the C-G program was an intricate system of surveillance and control over entire communities and a tool to extract target intelligence on VC political cadres, guerilla formations, and various locations from individual residents. In what can be seen as an early precursor to the more infamous Human Terrain teams used in Afghanistan and Iraq during the war on terror[63], the C-G program was a way to link development,

104 Chapter 4

security, and killing through "anthropological" methods, in the belief that this could somehow provide insights that would lead to victory.

In addition to the C-G program, interrogation of captured suspected Viet Cong affiliates served as another important element in the recurrent targeting cycles the Phoenix program was configured to become. In addition to the pure interrogation function of the PICs, Carver explains,

> each of the PICs also has a collation section into which interrogation reports are deposited along with intelligence reports from other Vietnamese intelligence organs. Biographic cards are filed alphabetically by hamlet, village, district and province as are reports broken down by specific topics such as 'VC Taxation,' 'VC Security Apparatus,' etc. The biographic material is readily retrievable for the preparation of 'black lists' of identified VC to support military and pacification operations. The cards are cross-indexed to individual biographic interrogation reports giving additional details.[64]

In a similar vein to the Bombing Encyclopedia, these blacklists served as a way not only to single out specific targets for neutralization but also to structure a particular understanding of how the VCI was organized as a hierarchical network and how it worked. Unlike the Bombing Encyclopedia, however, this archive of believed VC operatives was to be a much more fluid list that would be constantly updated as new data were produced and analyzed.

The reorganization of the counterinsurgency efforts in Vietnam in December 1967 was in large part due to the lack of coordination and collation between the above-mentioned collection programs as well as problems linking them to the broader efforts by MAVC and the CICV. The CIVC, although mostly concerned with OB, had created a Political Order of Battle Section aimed at identifying and tracking the VCI that also produced an archive of dossiers on suspected VCI members that could then be recalled by an automated system.[65] While each of the collection efforts created "a laminate of the VC organization," the problem was "not so much the acquisition of new data as the collation and exploitation of the information" being obtained through these collection sources.[66] In order to amend this, the Infrastructure Intelligence Coordination and Exploitation Structure (ICEX) was created in July 1967.

It was a "new staff structure ... designed to bring all the foregoing programs, as well as a number of MACV programs, under the operational control of Deputy to COMUSMACV for CORDS (Ambassador Komer) and into an integrated and sharply-focused attack on the VC in order to achieve unified line of command and a sharp stimulation of anti-infrastructure operations."[67] Coordination of intelligence, as its name suggests, was the program's paramount objective. Although decentralization of collection and dissemination was key to ICEX, the analysis would be centralized and then fed back to the districts.[68] The establishment of ICEX institutionalized manhunting in the Vietnam War, fusing intelligence and operations and the different disparate elements into a coordinated assemblage to go after the VCI.

When the Phoenix program was launched at the end of 1967, it built a vast network of over 100 provincial and district operations committees in South Vietnam, whose essential work was to create lists of "known" VCI operatives, their affiliation and rank within the network, and possible location. Once all this was compiled, the operational arm of Phoenix, usually local militias such as the PRUs or police forces, would kill or capture the suspects and interrogate them for more information on the VCI, which subsequently would keep the targeting cycle in motion. Crucially, operations that emanated from these blacklists were in large part designed to capture, not kill, the suspects. A CIA fact sheet on the Phoenix program, dated November 25, 1969, stated that "defection or capture are the preferred methods of neutralization as the individuals often provide highly useful information which leads to additional neutralizations."[69] As John Mullins, an American PRU advisor put it, "prisoner snatches were key. You can't get information out of a dead man."[70] A dead VCI suspect would mean an intelligence dead-end was the mantra.

This is by no means to deny that killings were often the result and probably the intention of many Phoenix operations, but critics of the Phoenix program[71] have often overlooked this crucial aspect to the manhunting in Vietnam and its continuities in today's global targeting world. To keep the targeting process running, capture and interrogation, often using severe torture, became vital elements in the fusing of operations and intelligence into a seamless whole in which they fed off each other.

The Phoenix program was a culmination of several efforts to transition from the widespread military-focused "cordon and search" or "search and destroy" approach to neutralizing the VCI to the "specific targeting" of individual VCI suspects. Robert Komer described it as "analogous to a 'rifle shot' rather than a 'shotgun' approach. Instead of cordon and search operations, it will stress quick reaction operations aimed at individual cadre or at most small groups."[72] This new practice, *specific targeting*, would, according to Colonel William Knapp, need "to use modern, skilled, police techniques to go after individual VCI," and he added that that "specific targeting requires trained and dedicated personnel, and well organized data bases to be effective."[73] Even though province/district intelligence and operations

106   Chapter 4

coordinating centers (PIOCCs/DIOCCs) had been set up to facilitate this new method of operation, it was, according to Knapp, not without its difficulties and problems. Troubles with information sharing and coordination, among the Americans themselves and between the Americans and the South Vietnamese, would continue throughout the program's existence.

Despite these problems, the program continued full steam ahead, with the aim of producing another level of granularity to the view of below, down to the smallest unit: the individual and its functional relationship to the larger VCI network. The way in which its epistemic operations worked, however, would see to it that this turned into an indiscriminate system, constantly, to paraphrase Ian Hacking, making up the wrong people, sweeping up individuals, and killing people at a frantic pace.

Reconfiguring the Enemy

According to the Phoenix advisors' handbook—the manual used to train US advisors on the Phoenix program—the US saw the VIC as "the political and administrative organization [or] 'shadow government' . . . through which the Viet Cong control or seek control over the South Vietnamese people . . . trying to provide a viable alternative to the GVN."[74]

The central effort of the program, the handbook explains, is to place all the VCI intelligence gathered by all the different agencies in one location and systematically collate anti-VCI operations based on the Phung Hoang Center's data bank. P/DIOCCs would coordinate collection and collation of intelligence and operations, serving as fusion centers for constructing targets and producing enemy models. The operational procedures for the Phoenix program and the construction of targets should, according to the handbook, work in the following step-by-step way:

> Once a suspect VCI is identified (name and VCI position are known), two index cards are prepared and catalogued both in alphabetical and village/hamlet files (pgs 79–81), The next step is to develop a VCI Target Folder on the individual; a good target folder will enable the cadre to be specifically targeted (you will know his habits, contacts, schedule, and modus operandi) . . . Once a target folder is established, . . . other files and charts such as source control records (pgs 73–75), VCI organization charts (pgs 142–144) photographic files (pgs 97, 98) and files on guides (pg 98); will each in their own way tend to sophisticate the process of specific targeting of VCI cadres. Consistently up-dated organization charts will particularly aid in the development of operations to neutralize entire VCI staff elements or organizational echelons. An IOCC [Intelligence Operations

and Coordination Center] is not functioning efficiently unless it regularly and successfully targets specific VCI cadres and organizations for neutralization.[75]

The handbook pays attention to the "careful, professional preparation of VCI Target Folders containing, both the 'VCI Target Personality Data Form' and the 'Offender Dossier' for each and every member of the VCI throughout the country [which] is the, foundation from which successful operations can be run." It continues in its bureaucratic manner: "The folders must contain all the, available data on an individual and be constantly reviewed and updated. . . . Copies of source reports, captured documents, interrogation reports, Hoi Chanh debriefing reports, and all other relevant documents are to be filed in this folder so that all information on a specific target is in one place where it can be reviewed easily and quickly."[76] In Annex A, *Operational Planning Guide*, the handbook goes into further detail on how to go about collecting and analyzing information on the VCI, ranging from the general to more detailed analysis of VCI patterns of activity and how to plan and execute an operation and exploit the intelligence from the operations and add the intelligence to the "Local Data Bank."[77]

These operational procedures for the Phoenix program show the remarkable bureaucracy and command and control that went into the system and the sanitizing language of the program. It clearly describes the bureaucratic end-to-end operational process of creating a target using multiple intelligence sources and methods, how this target fits into and updates the VCI organizational network, and the way in which this system is designed to feed on and digests information generated through feedback operations.

The advisor handbook is keen to stress that these "files are only a tool by which to accomplish the mission,"[78] but the way in which the system was configured and operationalized meant that these files became more than just tools. They came to drive not only the Phoenix program and its operations but, increasingly, how the VCI was produced, conceptualized, and operationalized. The files served as the layers through which the enemy was understood and operations planned and conducted.

While the HES created a "national mosaic composed of individual hamlets,"[79] the Phoenix can be said to have created a mosaic of the VCI composed of individual humans. In this way, the enemy was produced as nodes in a complex relational network distributed across the sociopolitical environment. Apart from the obvious difference in what targets were, relationally linked individuals rather than causally connected individual infrastructures of the BE, network thinking introduced methods and logics that differed from systems thinking. In many ways, Hunter Heyck argues, network thinking emerged as a correction to the more static, hierarchical, and rigid systems analysis, which was often seen as being unable to discern more complex social structures.[80] Not surprisingly, this was something that militaries and other security agencies picked up and utilized in the belief that it would identify functional roles, organizational positions, and influential individuals.

The favored tool for analyzing networks was *social network analysis*, or SNA, a conceptual and methodological tool that was originally developed by social anthropologists and designed for mapping social dynamics and capturing patterned social relations within societies.[81]

Armed with methods like contact-chaining and pattern-of-life analysis, as we call them today, the belief was that by linking suspicious individuals, based on their habits, contacts, schedule, and modus operandi as the above operational procedures of the Phoenix program outlines, and their social relations to each other, a presumed network structure and its power relations would emerge. Because networks, unlike systems, are structures without logical limits, the result was an ever-expanding network in which more and more people were deemed suspicious, based on their relations to other individual elements, or simply their behavior, and added to the files.[82] In principle, anyone could be a VCI if he or she had one time been in contact with someone who was already named or suspected to be a VCI.

Unlike the BE and the enemy-as-a-system modeling, which favored formal instrumental knowledge with material and logical limits to what could and should be included in the system, the Phoenix program was reliant on and sought out contextual and situated knowledge with few limits on inclusion. The result of this was a proliferating mosaic stored in an active database that would not work "efficiently unless it regularly and successfully target[ed] specific VCI cadres and organizations for neutralization."[83]

The Enemy in the Database

Central to this network mosaic was the VCI Neutralization and Identification Information System (VCINIIS), later renamed the Phung Hoang Management Information System (PHMIS). This was a management information system that generated a series of reports providing information on operational results as well as demographic, ethnographic, and biographical data on the VCI.[84] According to Douglas Valentine, this data infrastructure

> climaxed a process begun in February 1966, when Secretary of Defense Robert McNamara established the Defense Department's Southeast Asia Programs Division. The process was carried forward in Saigon in January 1967, when the Combined Intelligence Staff fed the names of three

thousand VCI (assembled by hand at area coverage desks) into the IBM 1401 computer at the Combined Intelligence Center's political order of battle section. At that point the era of the computerized blacklist began.[85]

But this system was more than a computerized blacklist of people to assassinate. This process served multiple purposes and various functions, although no less violent, for finding and killing/capturing individuals, for measuring progress and as management systems, and for modeling and simulating the enemy. According to the Command Manual for PHMIS, the system had three objectives: maintain the files containing biographic data on identified members of the VCI, including their job or position within the infrastructure, provide accurate statistical information on the neutralization of the VCI, and enhance the capability of the Phung Hoang/Phoenix program to respond to inquiries by interrogating the database, limiting the need for manual extraction and aggregation of data.[86]

The PHMIS was a reconfiguration of the VCI Neutralization Identification Information System (VCINIIS), which was a simple system for producing statistical reports on neutralized VCI members. This system, however, could not cope with policy changes that made it paramount not to index a captured individual as "neutralized until his final disposition could be determined." Thus, the command manual states, "It became evident that a simple statistical gathering concept was evolving into a system that would perform certain tracking functions."[87] The VCINIIS could not, however, track a presumed "at large" cadre, and the PHMIS was developed. The system itself operated through a number of information flows, processing various inputs into outputs that again served as rudimentary feedback loops into the system. Although the PHMIS operated on state-of-the-art computer equipment, human operators were very much part of its highly bureaucratized and standardized input-output operations, producing computerized blacklists:

> As a VCI cadre is identified, the Phung Hoang Committee assigns him a VCI number and sends all available information to PHD [Phung Hoang Directorate].... After these reports are received at PHD, they are reviewed for legibility and reasonability. The reports are then sent to National Police Command Data Management Center (NPC/DMC) for keypunching. The keypunched cards are sent to the MAC Data Management Agency (DMA) for processing. The first step in the system is processing by an edit program which lists erroneous cards and prints a message pointing out the error. [After corrections are made] the PHMIS master file is updated and all reports for the month produced.[88] ... The

output reports are sent back to PHD for distribution to each province and district.[89]

The PHMIS received two different types of input. One was neutralization information that was entered when a VCI suspect was killed, captured, or defected to the South. The other was the detailed records form or input sheet containing biographic information on a suspected VCI cadre. These sheets and their relation to other biographical material in the database could be updated as more and more information on the VCI was obtained.[90]

These report forms (figure 4.2[91]), with an emphasis on biographical and locational information and position within the VCI, looked very much like the Basic Encyclopedia target cards encountered in the previous chapter. They were a highly stylistic and managerial way to represent a human being, whose life was translated onto this piece of paper, and once it was abstracted, was placed on the kill/capture list.

Together, these forms of all confirmed and suspected VCI members made up the PHMIS master file, which analysts could access and cross-reference. Through this process, "a list of identified VCI is produced and used by the district committees for targetting [sic]."[92] In addition to this computerized backlist, PHMIS produced other outputs, such as short statistical reports on numbers of VCI neutralized, but the ability of the database to easily and speedily create organizational charts of the VCI network using available inputs was crucial to the way the enemy came to be produced.

The VCI Organizational Profile Report, as it was known, was the "basic operational support document of the PHMIS," through which "the entire VCI structure as identified [could] be displayed on a computer printout."[93] The Command Manual calls this an "extremely valuable operational tool" used to cross-check information, in order to "bring a *semblance of order* to the VCI data holdings" and "promote orderly and professional tasking of intelligence collection organizations and effective specific targeting."[94] As a database, the PHMIS produced inherently instrumental information based on the arrangement and organization of a data model that allegedly related the various bits of data to each other logically. The data logics and management techniques used to produce the technical artefact, or target, were thus of great importance since the way the data model was manipulated meant that it produced novel visions of order.[95]

A rarely discussed but crucial aspect of this database was the built-in ability to make queries; "PHMIS has the capability to produce one-time reports in response to queries with minimal programming effort. This is done through a subsystem called the Phung Hoang Inquiry Response Module

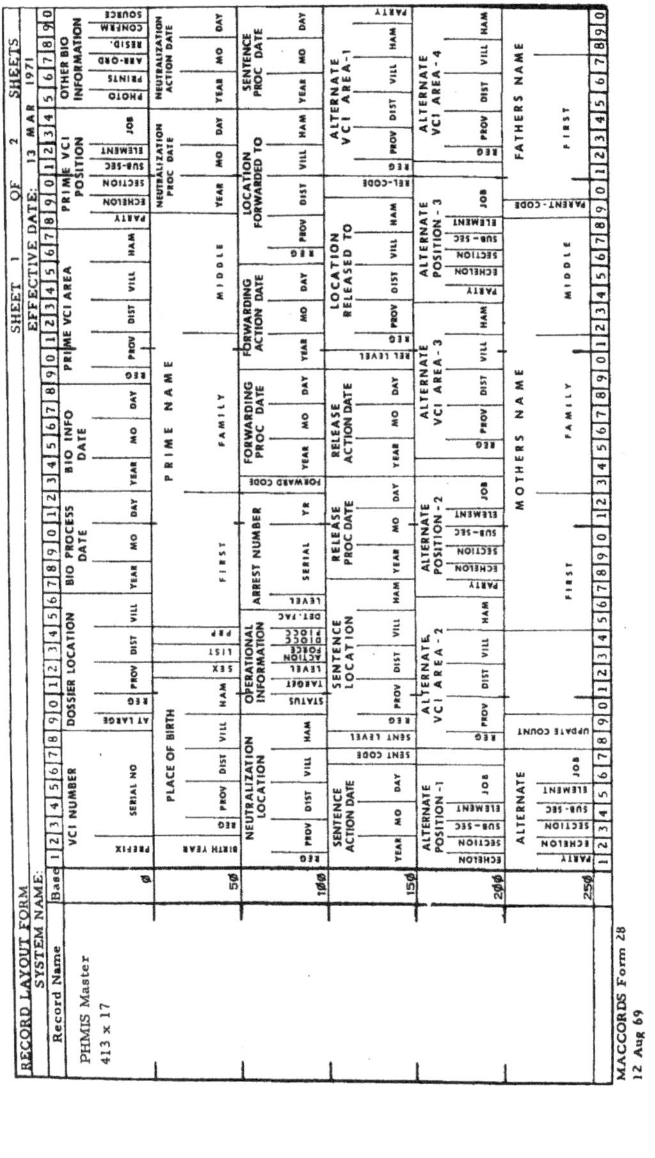

Figure 4.2

PHMIS report form. *Source*: US Military Assistance
Command Vietnam (MACV), Command Manual,
Phung Hoang Management Information System
(PHMIS), 1969–1972, Headquarters, U.S. Military
Assistance Command, Vietnam (Saigon, 1972):
Appendix G 2.

(PHIRM)," which "allows the user to select data record[s] from the PHMIS data base by specifying a condition or a series of conditions that the record must meet to qualify for extraction."[96] These records were then ordered according to "user specified sequence," producing sets of parameter cards that could be "matched against the PHMIS data base in a single computer production run."[97] A sample figure shows that these runs could include, for instance, "distribution of VCI cadre identified during 1970 who were more greater than 50 years old."[98] The PHMIS, being part of the aforementioned NIPS system, gave users the ability to structure files, generate and maintain files, revise and update data, select and retrieve data, and generate reports. In some ways, "NIPS supported relational database functionality."[99]

What is crucial here is that the PHMIS's ability to function as a relational database transforms the system from being a means of organizing, storing, and retrieving structured data, as with hierarchical databases, into a search device and a means of querying. This enabled the PHMIS to explore the data in ways not immediately apparent to analysts and search for relations and associations and follow hunches, making it possible to presumably produce "novel" insights from data.[100] While the HEW and the PHMIS report forms were designed to reduce human cognitive wandering and subjectivity through standardized formats, the ability to query the PHMIS opened up and allowed for a different type of epistemic exploration and violent experimentation based on the configuration of humans, machines, and the environment.

Although these novel and intricate infrastructures and practices for managing and processing data were important elements in the Phoenix assemblage, it was perhaps the rebranded counterterror teams, the provincial reconstruction units (PRUs), who provided the most effective anti-VCI actions, measured in the favored way, via neutralized VCI cadres.

The PRUs were the action arm of the Phoenix program. Organized, trained, equipped, and funded by the CIA, each PRU consisted of a team of eighteen locally recruited men along with an American advisor. At their height, there were more than 5,000 PRUs. Although they were the action arm, the PRUs were more akin to "an intelligence driven police force" whose paramount mission was not to kill, but to capture VCI suspects and exploit them for further intelligence. Being locally recruited, they allegedly had "intimate and complete knowledge of the people and terrain [leading] to a great ability to develop accurate intelligence on the VCI and to plan methodologically."[101] In other words, they had information about who was deemed suspicious by the local community.

Based on known suspects, the PRUs had developed their own low-tech targeting cycle for manhunting. Starting off with a "known" suspect, they would seek out the individual's social relations to add individual elements to the network and determine suspiciousness based on certain traits of behavior. For instance, as Andrew Finlayson, an American advisor in Vietnam, explained, "seventy-five percent of the time, the PRUs did their own targeting: 'This guy's sister is pro-VC, he comes to the market and is buying way too much food,' etc."[102] Such assumptions would serve both as a way to determine suspiciousness and as a clue to be further examined.

Utilizing their own forms of signature hunting based on such forms of contact-chaining and suspicious pattern-of-life behavior and exploitation of captured individuals and materiel, the PRU intelligence system and operations became invaluable to the Phoenix program. By spring 1970, when the Census-Grievance program had died out, Thomas Ahern notes, "The PRU had become the primary collector as well as the 'only effective Phung Hoang operational asset.'"[103]

For all its automated data processing, storage, and automated queries, Phoenix was still reliant on local people to turn in cadres and on human sensors and operators to do the input work. It was a highly subjective process reliant on contextual and situated knowledge, but as with HES it was cloaked in objectivity, as the information passed through the automated computer systems. As Vic Croizat, a US Marine colonel turned RAND consultant, later reflected, "A profile [of a person] is not an objective thing. Somebody's got to draw it up" and, to his mind, "no matter how honest you are when you do these things, there's got to be an element of uncertainty . . . from a mere interpretation of 'various bits of evidence,' people could be dealt with in a 'definitive way.'"[104]

While the number of neutralized VCI cadres points toward a highly successful program, the validity of these numbers is highly questionable, with a large part of those neutralized being based on uncertain identifications rather than real VCI cadres. Partly because of the perpetuating and ever-expanding logics inherent in network thinking and the newfound targeting cycle where military operations would feed intelligence and vice versa, but also because of managerial incentives to produce numbers, the Phoenix program led to the sweeping up of large numbers of innocent people. The obsession with statistical metrics meant that for the "first 2 years of Phoenix, each province was given a monthly quota of VC to neutralize, depending on the size of the infrastructure in the province," regardless of there being any known cadres present in that region.[105] All in all, the Phoenix program is perhaps best summed up by then–US Secretary of the Army Stanley Resor, who described Phoenix as "an indiscriminate 'dragnet method' of attacking the VCI [ignoring] the 'social and moral costs which that might entail.'"[106]

Costs included not only those thousands of innocent individuals caught up in this web, but also the shattered families, villages, and hamlets that were viciously trapped in this accelerated military targeting cycle.

Legacies of Intelligence and Operations Fusion

The Phoenix program was in essence an ecology of operations aimed at mapping, understanding, and dismantling the presumed ecosystem that sustained the insurgency in South Vietnam through intricate epistemic operations linking information in context to the overall goals of the war-fighting machinery. Armed with an IBM 360 model 501 computer running on a standard IBM 360 operating system (OS), with all source programs written in ANSI COBOL and supported by standard punch and card verifier equipment—a card sorter, a card reader/punch, an IBM 2314 disk unit, five IBM 2400–3 tape drives, a line printer, and a CalComp plotter—the Phoenix program produced its enemy through the collaging of "various bits of evidence."[107]

Through contact-chaining and pattern-of-life analysis of known suspects based on human sensor inputs from the field, the automated processing of the data slowly conveyed an intricate network of nodes that would become known as the VCI. This model was not merely metaphorical representations of danger or an objective threat but the result of epistemic operations classified and filtered out of what could be seen, effectively translating the world into actionable entities.

This network logic, when designed and engineered into the cyclical targeting assemblage of Phoenix, produced sociotechnical practices that were not standard processes of objectification in which persons become bodies but, rather, were the other way around. Bodies and their personhood and behavior became the objects of targeting based on predefined signatures of suspicion (e.g., "This guy is buying way too much food."). This is a particular form of classification in which it is not the person that is targeted but, rather, their personhood. The result was a reconstruction of the battlefield toward social structures and relations that encouraged the further blurring of civilian and military spaces, leaving in their wake a vast trail of dead bodies, counted and statistically measured, and devastated social structures.

Phoenix's translation of the environment did not produce a narrative of the conflict like the HES but, rather, a computerized collection of individual relationally linked items that constructed a distinct instrumental view of the enemy. Crucial to this representation was the database, producing its own form of knowledge in which the enemy became known through the configurations of the system and its intricate epistemic operations between humans, machines, and the environment. Although the Phoenix program was designed and engineered to centralize the management of manhunts in Vietnam through computerized processing of information and distribution, it drove forward a vision of the enemy and a way of war that was in large part constructed through the cyclical targeting methodologies and the inner logics of the data infrastructure and its input-output processing.

While many critics have pointed out that the Phoenix program was no more than a computerized assassination program,[108] the true value of this program for the military was, however, not only in its ability to kill but also in its design as a learning instrument; that is, through cyclical operations new data would be produced that they believed could be exploited by churning and collaging bits and pieces through the computerized system. The assumption was that this process would lead to further knowledge of the enemy network, generating new targets and operations. This learning was, of course, situated in the sense that it was a function of the context in which the ecology of operations was working.

Fusing intelligence and operations into a seemingly seamless whole, the Phoenix program had designed and engineered an action-oriented cyclical targeting process that reinforced the need for constant operations to learn about the enemy. The result was a reduction of the war into discrete tactically oriented targeting operations aimed at feeding the system, creating its own cyclical logics, with perpetuating military operations serving as an end in itself. This is a practice, which we will see in the next chapter, that takes on a whole new dimension as it is entangled with so-called big data and advanced computational analytics.

While the whole Vietnam War proved to be a disastrous endeavor for the Americans, and an outright catastrophe for the Vietnamese people, the Phoenix program prefigured the methodologies and epistemic infrastructures and practices for manhunts in the years to come through its intelligence-operations fusion. Although out of the operational hands of the military for the foreseeable future, the Phoenix program arose from its ashes elsewhere: in the Philippines, El Salvador, Colombia, and other Latin and South American countries through various CIA-initiated operations in the 1980s.

The military had also "doctrinized" its lessons. The 1973 version of the field manual, FM 30-5, *Combat Intelligence*, which had up to that point been mostly interested in battle-of-order intelligence, contains a section in the last chapter euphemistically called "Stability Operations." Here, it is acknowledged that "intelligence data base requirements [for counterinsurgency] differ greatly from those needed for conventional military operations. Stability operations require elaborately detailed intelligence concerning *sociological, political, geographic, and economic information.*"[109] Dividing the need for tactical (whereabouts of guerilla elements) and nontactical (human terrain) data, the FM 30-5 had now doctrinized and institutionalized a view of below that included not only hamlets but also individuals, the smallest possible unit of analysis.

The Worms

In 1968 the American military launched Operation Igloo White, seeding the jungle highlands in Laos with a network of seismic, olfactory, and auditory sensors that gathered sensory events and relayed signals to data-processing facilities in Thailand, where the information could be processed to map movement in real time.

Igloo White was the technical solution to the problem of the Ho Chi Minh Trail, a vast supply-chain network of trails and tracks that had been constructed by the North to support their war effort in the South with a continuous supply of personnel and materiel. The US had previously and secretly been flying interdiction missions into these areas to bomb the trails, but these had not, mainly due to the inability of the Air Force to find targets in the dense jungle terrains, yielded any measurable effect on Hanoi's ability to sustain the war in the South.

To stop this flow of weapons, supplies, and troops into the South, McNamara and his confidantes sought, as they usually did, a technical solution to the problem. After a meeting in early 1966 with members of the infamous Jason Group,[110] McNamara formally requested that the scientists look into the feasibility of "a fence across the infiltration trails, warning systems, reconnaissance (especially night) methods, night vision devices, defoliation techniques and area denial weapons."[111] With its 1966 summer study, the Jason Group came up with a plan to create a barrier stretching from the South China Sea, south of the DMZ, across the Laotian frontier, to the border of Thailand. The original plan consisted of two barriers; one physical barrier of mines and barbed wire and the other, an air-supported electronic barrier using various electronic sensors. In mid-1968, however, the barrier concept was reduced to an aerial, sensor-based electronic interdiction program that was to be operationalized throughout Laos.

Although it goes by many names—the electronic wall, the McNamara line, or simply the electronic battlefield—its codename was Igloo White, and it "was truly the start of the high-tech sensor warfare that continues to this day."[112] The idea was simple but technically complex and consisted of a tripartite operation producing, processing, and making actionable data on the environment: first, the known or suspected trails were seeded with acoustic, seismic, and olfactory sensors, dropped mostly by navy and air force planes, to detect troop movements and truck vibrations. Those sensors would broadcast to an orbiting airplane that relayed the signals to the Infiltration Surveillance Center (ISC) at Nakhon Phanom Air Base in Thailand. Second, the ISC would subsequently analyze the signals to produce tactical information for planning and interdiction operations providing real-time targeting information for tactical airstrikes. The moment the sensors picked up "noise," analysis would be conducted, and finally, air strikes could be launched against the sensory location.

Although simple in its conception, Igloo White was a creative invention utilizing state-of-the-art technologies and novel human-machine configurations to create a remote networked sensor-to-shooter system, supposedly making the terrain computable and translating the enemy into an anonymous technical signature akin to a submarine—locatable, trackable, and shootable. Not only did this targeting system, as Derek Gregory argues, "prefigure the technical infrastructure for today's drone wars,"[113] but it configured a very distinct ecology of operations reconfiguring the role of human perception and cognition in concert with that of the enemy in which both became configured as nodes in this experimental assemblage of destruction.[114]

Designing and Engineering Igloo White—from Sensors to Shooters

At the heart of the Igloo White system were the extensive strings of electronic ground sensors that picked up and transmitted audio, seismic, and olfactory indications. The early audio sensors were off the shelf, but these were gradually modified into the canopy acouboy (designed to hang in jungle canopy) and the spike acouboy (to be implanted in the ground). These acouboys were later augmented in application by seismic sensors, such as the ADSID, and, for some applications, radio frequency, infrared to pick up heat signals, and magnetic sensors to detect weapons. One of the more experimental sensors used in Vietnam was the XM-3 airborne personnel detector, also known as the "people sniffer," which used a detection method dependent on effluents unique to human beings, such as urine and sweat. In the end, the sensors served as human sensing prosthetics.

Decisions on where to plant the strings of sensors were made by the Sensor Placement Planning Committee at Task Force Alpha (TFA), a Thirteenth Air Force organization under operational control of the Seventh Air Force. The integral roles of the TFA were monitoring, interpreting, and maintaining the Igloo White sensors.[115] Studies were made to determine their feasibility, taking into account such factors as terrain, jungle canopy, tone codes and frequencies available, and the type of information desired from the string. This consisted of a network of some 20,000 sensors that were dropped in strings of five or six. As a tactical evaluation of the program from 1970 explains, "All sensor drops were accomplished during daylight when weather conditions made ground photography possible. This was necessary since a KB-18 camera took horizon-to-horizon photos at the instant of sensor release. Using this photo coverage and established ballistics statistics, the TFA personnel were able to compute, within about 40–60 meters, the exact location of each sensor."[116]

These coordinates were crucial for tactical precision airstrikes, but given the relatively low accuracy of the computation (within forty to sixty meters), different types of cluster munitions, such as the CBU-55 and CBU-72 fuel-air incendiary bombs or the CBU-75 Sadeye, were the ordnance of choice, literally turning the jungle into a burning inferno through which thousands of razor-sharp shards came flying.

However, the sensors could not function without the other vital elements of this system. Due to the size and complexity of the sensor fields themselves, vast computational power was required to integrate all the sensory input and ensure proper analysis and data readout. But before the sensor output could reach the ISC in Thailand, it had to be relayed via orbiting aircraft that received signals from the sensors, amplified them, and retransmitted them to the ISC. "When activated by movement or sound, the sensor would transmit its basic identity code to an EC-121 aircraft which flew a specified orbit above the sensor field. This aircraft would automatically relay these transmissions to the ISC for analysis."[117] Having finally reached their goal, the signals would be processed at the ISC. The center performed three functions: data processing, target identification, and system performance monitoring. This man-machine-environment configuration worked thusly:

> The digital data picked up from the relay aircraft were fed into the Ground Terminal System Segment. Audio information was separated from the data train, converted into analog form, and sent to audio monitoring specialists who would make both an audio and spectrum analysis of chosen audio input. Any selected audio information could be entered into the computer using a 2260 video display-typewriter. Tone code information was fed directly into the IBM 360/65 computer. There it was combined with the audio assessment and printed every five minutes in the form of a hard copy printout. This record covered the minute-by-minute activations of each sensor for the previous 40 minutes, providing a visual history of activations. In addition to the hard-copy printout, each one minute update of the CONFIRM was displayed on a 2250 cathode ray tube which provided near-real time tactical information.[118]

There were numerous problems with the sensors, at least in the early years. Some did not survive the drop, but if they did, they ran on batteries that lasted only a few weeks and then had to be replaced, either from the air or from the ground. The biggest problem with the sensors, however, was the number of false positives they created. These were triggered either deliberately by the North Vietnamese as they learned how the sensors worked or simply by animals, aircraft overflights, or even heavy rainfall. This would, however, we are told, be amended through the use of computer algorithms, implemented by two advanced IBM-360 computers that enabled the system to filter out noise, and by "human expert judgement."[119]

To validate the information from the sensors, and to avoid false positives, the ground surveillance monitor (GSM) used the number of sensor activations, their strength and duration, as well as their movement from sensor to senor to assess the amount of traffic and its size, speed, and direction. Added to this was intimate and specialized operator knowledge of the area, the location of the sensor strings and the individual characteristics of the different sensors.[120] It was the twenty-four-hour combat operations center (COC) that continually monitored sensor activations, and the so-called most promising detections were passed to the Airborne Battlefield Command and Control Center (ABCCC) as SPOTLIGHT reports for possible strike, which then allocated air assets for attack if and when required. The targets were determined and prioritized by the TFA.[121]

Igloo White was highly reliant on all three technical components of the system—sensors, relay aircraft, and computers—but also on human operators. A milestone in the development of the Igloo White system, we are told, nevertheless came with the conversion from IBM 360 Model 40 computers to Model 65. This greatly increased the data-processing capability of the TFA, allegedly enabling the detection, tracking, and analysis of real-time tactical information, integrating intelligence, and furnishing real-time direction of airstrikes on moving targets. Under the Commando Hunt/Bolt programs, the system was configured in such a way that it could track a moving target through sensor-by-sensor analysis. If something triggered so-called sensor events along a sensor string, one could now infer from this its speed of movement and its future location. As such, one could, in theory, direct strikes at a predetermined geographical location based on estimated time of arrival for the entity that triggered the sensors.[122]

These forward air control operations, code-named Sparky, were controlled by a three-man team sitting in a room; the forward air controller (FAC), the radio operator, and the sensor interpreter (SI). In front of the FAC and the SI was a workstation consisting of an IBM 2250 monitor, on which several displays could be selected, tied to a computer. This human-machine team used the sensor events to calculate the speed and trajectory of the moving "worm" target to define predetermined strike points (figure 4.3[123]). The whole operation worked thusly:

> As sensor activations were displayed on the 2250, the interpreter and FAC conferred to determine the validity of an activation. They might cross-check it with other sensors or have one of the sensor audio monitors listen to the sensor for further information. If they determined the activations were a valid target, they changed the presentation on either or both of the tubes to the map presentation of the sensor string in question. DMPIs [desired mean points of impact] were displayed on the map, which featured a *"worm"* which moved down the map at a rate equal to the computed target speed. ETAs [estimated times of arrival] for various DMPIs were also visually displayed and constantly updated. The sensor interpreter and FAC were thus able to "see" the movement of the truck and determine the time for a strike on one of the DMPIs.[124]

For the first time, observing moving targets by sensor interpretation was made possible, and airstrikes could be directed on a near-real time basis using a "blind-bombing" method. The coordinates for the DMPI would be plotted into the strike aircraft (e.g., F4s) computers, which would then automatically release the wide-area ordnance once the aircraft was in strike position, obliterating anything within range of the CBUs.[125]

This system was, of course, reliant on the assumption that anything that moved within the area of the sensors was a worm. It did not have the ability to, nor perhaps an interest in, distinguishing between friend and enemy, civilian and combatant. Sensors, acting as prosthetics for human senses, were blind insofar as they could detect only predefined audio, movement, heat, or smell signatures in a selected area seeded with sensors. The system had no means of verifying enemy combatants—only the capability to monitor a known location for a known signature. Thus, in this system, anything that moved within the parameters of the McNamara line, triggering a sensor event, was by definition an enemy and was subsequently turned into a worm target— and being a target in this ecology warranted a CBU from above.

As Derek Gregory notes, the Igloo White, like many other bombing systems, "works to render bombing an abstract, purely technical exercise for those who execute it."[126] The Igloo White system was thus not interested in mapping and making visible this terrain for the sake of creating knowledge; it only monitored the environment and made it computable for the sole purpose of enabling real-time action within this technically abstracted

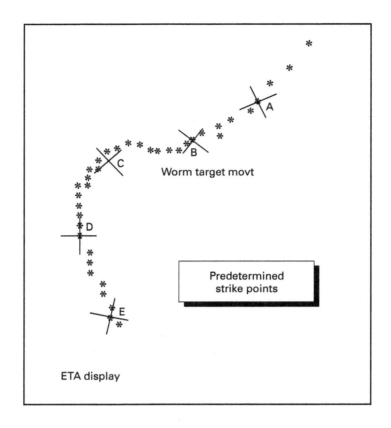

Figure 4.3
A worm and its predetermined strike points. *Source*: Philip D. Caine, Igloo White, July 1968–December 1969 (U) ([Hickam Air Force Base, Hawaii]: HQ PACAF, Directorate, Tactical Evaluation, CHECO Division, January 10, 1970): between 16 and 17.

context. This was a new form of territorial construction, producing a novel operational environment in which anything that moved was a target.

Igloo White was highly experimental. Tests and trials were done on sensors, and the resulting data were analyzed, constantly looking for the best configurations to provide the most sensitivity compatible with reliability. For all the problems and difficulties with the Igloo White system—implant loss, unreliable signals and low battery life on the sensors, data loop loss within the relay system caused by radio interference, equipment problems on the relay aircraft, and the occasional overloading of the equipment at the ISC—the overall conclusion was that the system worked well; "sensors saved U.S. lives"[127] was the mantra, shifting the burden of dying to the Vietnamese.

Despite this conclusion, Seymour Deitchman, one of the founding figures of the electronic battlefield, lamented the failure of the system to stop *all* truck transport along the trails in Laos. This, he argues, was due to a failure to implement the system as designed, or at least as envisioned by him and his colleagues within the Jason Group. He believed that "the conversion of the intended tightly integrated reconnaissance-strike function to simple surveillance, with strike aircraft called in as available—compromised the system's ability to do the job for which it had been designed." But, he contended, the effort might have been pointless anyway. Due to the size of the area and the effort the North put into the war and its supply chains, there might not have been enough resources for it to work as designed.[128] Regardless of the question of whether or not the system would have worked, the abstraction of the environment and the rendering of the process as a purely technical exercise made Igloo White into a devastating project for producing and controlling the jungle highlands of Laos and Cambodia.

Prefiguring the Future

Regardless of its inability to win the war for the Americans, this electronic battlefield had shown, to its proponents, that it was possible to track and strike a moving target using sensor interpretation on land. It was famously heralded by General William Westmoreland as the battlefield of the future:

> On the battlefield of the future, enemy forces will be located, tracked, and targeted almost instantaneously through the use of data links, computer assisted intelligence evaluation, and automated fire control. With first round kill probabilities approaching certainty, and with surveillance devices that can continually track the enemy, the need for large forces to fix the opposition physically will be less important . . . I see battlefields or combat areas that are under 24 hour real or near real time surveillance

of all types. I see battlefields on which we can destroy anything we locate through instant communications and the almost instantaneous application of highly lethal firepower. I see a continuing need for highly mobile combat forces to assist in fixing and destroying the enemy.... Our problem now is to further our knowledge—exploit our technology, and equally important—to incorporate all these devices into an integrated land combat system.[129]

The future was closer than Westmoreland realized. More and more resources were made available to see that the future battlefield was indeed electronic, transferring the burden of conducting war from people, that is US soldiers, to machines. By the early years of the 1970s many future applications for the system were recommended, including a worldwide border and interdiction system that would relay to satellites or drones rather than orbiting aircrafts, the US Navy's global Sound Surveillance System (SOSUS) and the US Army's Remotely Monitored Battlefield Sensor System (REMBASS).[130]

Merging sensors, computers, and shooters with humans and the environment, the Igloo White system set the Americans on course to develop novel military organizational and combat techniques that are now most famously described and doctrinized under the rubric of network-centric warfare as we saw in the last chapter.[131] The experimentation with drones as relay aircraft and the idea of replacing the manned bombers with unmanned ones certainly points in the direction of today's drone wars, but so does the practice of tracking moving targets by sensors.[132]

Of course, this type of tracking has a longer history with the development of radar for antiaircraft and antimissile systems, and sonar for antisubmarine warfare, and the more general development of cybernetic fire-control systems based on recurrent feedback loops. But this electronic battlefield was of a different kind, not only because of its connection to land warfare and its alleged ability to track humans but, more importantly, because Igloo White configured a special kind of human-machine-environment system. It did not simply replace humans with electronic devices for surveillance and shooters with remotely controlled bombers but reconfigured the roles of humans and the environment in the system.

The elements of the system—humans, sensors, computers, and the environment—configured an ecology of operations that fashioned continuous exchanges with each other through recursive chains. As we have seen, the human interacted with its environment and enemy through sensors, computers, and screens, constantly correcting and validating information based on specialized knowledge, recurrently feeding information back into the computers to update the environment. In this way, human perception and cognition was a vital part of signal processing and the electronic battlefield.[133]

The result of this configuration was that both operators and the enemy (targets) increasingly became configured as nodes in a complex network, distributing information flows across bodies, instruments, and the environment. The Igloo White system was thus an early example of what Lucy Suchman has termed the contemporary twin form of "deadly bioconvergence"[134] in which bodies and the environment are locked into a system of information processing, choice generation, and decision-making, distributed across a messy ecology of operation and a truly devastating assemblage of destruction.

While Igloo White was seen as the cure for the failings of the US military's conventional and counterinsurgency warfare in Vietnam, its true legacy was that it set in motion Westmoreland's dream for the future battlefield, creating the illusion and imaginary that war could be structured, fixed, managed, and won through computational control and automated knowledge processes and decision-making.

Conclusion

When the Nixon Administration took over in 1969 all the data on North Vietnam and on the United States was fed into a Pentagon computer—population, gross national product, manufacturing capability, number of tanks, ships, and aircraft, size of the armed forces, and the like. The computer was then asked, "When will we win?" It took only a moment to give the answer: "You won in 1964!"[135]

Harry G. Summers's reference to a story that was recited in the US Army during the later years of the war has become a classic for the many critics of the Vietnam War. To many, McNamara and his managerial approach to war with its computers, quantification, and business-like objectives is the cause of the US failing to win the war. Whether or not these "quantitative indicators" should be seen more as symptoms than the cause of the failings and utter destruction in Indochina is a moot question.[136] The analysis from this chapter, however, shows that it is not the managerial logics themselves but, rather, the operationalization and configuration of these imaginaries into various ecologies of operation that drove the thinking and conduct of the war. Various configurations did change the nature and the availability of information and knowledge for war managers, generating a model of quantitative indicators, trends, and targets that provided the war-fighting effort with technical fixes and best practice solutions. But, these were solutions that normalized killings, legitimized and justified destruction, and masked a horrific and devastating war effort as systematized progress.

For all their differences, each of the systems and programs—the HES, the Phoenix program, and the Igloo White system—were experimental attempts to make the environment computable and actionable. Through these configurations, information flows were collected, converted, analyzed, classified, and distributed across different instruments, operators, and their environments—each operating from different logics and rationalities, designed and engineered in different ways with different forms of human-machine-environment interactions, and with different analytical focuses at different levels of analysis: the population, individuals, and terrains. Seeking, in different ways, to put social and physical space under quantitative and computational control, the US martial apparatus sought to construct a battlefield that was visible and controllable, producing in its wake novel subjects, objects, targets, enemies, and operational environments and conditioning specific forms of violence.

The hope was in large part that the computer and the logic of data processing and analytics would create a sort of "mechanical objectivity"[137]—replacing the subjectivity and volition of individuals with routines of mechanical and computational reproduction—that could sort, categorize, and classify Vietnam to impose order, equilibrium, and stability to the active, fluid, messy, and unpredictable nature of human behavior and the social world. The result, however, was not a mechanical objectivity but, rather, a contingent ecology of operations with unpredictable developments, which had effects that were not instrumental but were generative and transformational, creating its own internal dynamics and largely operating as an autonomous force.

Here, it is important to be mindful of Bernard Geoghegan's comments in relation to technocracy. Technocracy, he argues, "is frequently misunderstood as governance by the means of technology and its related apparatus of data, bureaucrats, and nonexperts." However, he contends, the core feature of technocracy is, rather, the valorization of the technical as a neutral tool of governance in which conflict is seen as a mechanical failure that can be amended by technical experts. Thus, Geoghegan argues, "the essence of technocracy is not the technical as such but rather a political rhetoric of the technical."[138]

And the efforts in Vietnam can in large part be summarized as a belief in and valorization of the technical as a solution. Even though most of those running the war, from politicians down to advisors, expressed doubts and concerns regarding these systems, the HES, the Phoenix program, and the Igloo White system offered, in their minds, the best solutions to a problem they never understood or ever could from a technocratic worldview. While the war managers continued to rely on, valorize, and use these configurations for insights, they were in fact highly experimental and uncertain attempts to create informational feedback loops, rather than the neat and concerted effort to create managerial certainty and objectivity that is often portrayed. However, as this chapter has showed, it was not only the belief and the valorization of the technical that structured the course of the war. The technical itself, the contingent configurations of humans, machines, and the environment, created its own dynamics that often ran contrary to the technocratic vision of control.

For all their shortcomings, and despite the horrible human costs, the HES, the Phoenix program, and the Igloo White system showed the US martial apparatus the potential of configuring humans, computers, and the environment into a single entity, merging the view from above with the view of below, conjoining them with military operations in which learning and choice, not structures and system, emerged as the foundational preoccupation of information processing. So, despite all their failings, what these systems did in terms of targeting was in many ways to show the potential of marrying the two sciences of systems analysis and cybernetics beyond the laboratories of simulated nuclear warfare and into the real world. That the war in itself proved to be disastrous for the US, and even more so for Vietnam, did not figure into this martial calculation.

Indeed, it was not only the electronic battlefield that prefigured the epistemic and technical infrastructures and practices of contemporary warfare, but the idea of an overarching and all-encompassing system that could produce and process the entire environment, from the terrains to the population to individuals. This is a legacy that would truly revolutionize targeting and transform it into a violent learning mechanism in which everything and everyone is constantly targeted.

5

FEEDing the Network: Hunting for Signatures

It became clear to me and to many others that to defeat a networked enemy we had to become a network ourselves.[1]
— General Stanley McChrystal, Commander, US Joint Special Operations Command

Find-Fix-Finish-Exploit-Assess-Disseminate

After a quick and supposedly successful dismantling of the regimes in Afghanistan and Iraq, the US military soon came to realize that regime change did not equal victory. The enemy quickly dispersed, morphed, and resurfaced as a sustained insurgency aiming to oust the occupying forces. Ill prepared for this type of warfare, the US military had to reinvent past lessons and reconfigure its force to fight what they saw as a dispersed network of highly motivated, strategically wired, and continuously informed cells. The enemy, they came to conclude, seemingly forgetting the years in Vietnam, was "fundamentally different from any enemy the United States has previously known or faced."[2] Without fixed infrastructural targets or a "rigid—or targetable—chain of command,"[3] it was, much like the Viet Cong infrastructure (VCI), an enemy "whose primary strength [was] denying U.S. forces a target"[4]—an enemy that did not play to US strengths but, rather, camouflaged themselves in the terrains offered by nature or by everyday urban life. The solution, as US General Stanley McChrystal was later to describe, was to become more like them—hence the mantra "It takes a network to defeat a network."[5]

While drones and so-called targeted killings have received the bulk of the attention from media and scholars alike,[6] a handful of critical scholars, and the military themselves, credit the shift in operational practice from targeting fixed sites to hunting individuals to the reorganization and fusion of intelligence and operations. Steve Niva, for instance, argues that that the growth and evolution of the infamous Joint Special Operations Command (JSOC), which McChrystal oversaw, epitomizes the shift to a more networked form of organization and high-tempo violence within and across the US military.[7]

This was a transformation that came about through years of trial and error of operating under the banner of the "global war on terror." According to army intelligence officer Glen Voelz, today's counterterrorism (CT) network is "the culmination of a decade of tactical lessons, doctrinal adaptations, technical advances, and changes to the institutional cultures of the US military."[8] Bringing together different heterogenous elements, the CT network was a result of so-called lessons learned from Iraq and Afghanistan, experimentally combining network targeting, fusion of operations and intelligence, all-source intelligence, and key technological innovations in command, control, communications, computers, intelligence, surveillance, and reconnaissance (C4ISR).[9] While analysis of the bottom-up process of networking special operations forces (SOF) is a welcome contribution to critically analyzing contemporary warfare, networking, assembling, and configuring the military for manhunts is only part of the story. When you place military targeting at the center of this transformation, a different story emerges.

The early crucial aspect of reconfiguring the US martial apparatus into a global manhunting machine, was to fuse the previously separate fields of intelligence and operations. This fusion was designed from a logic that the unknown would only be found if the martial apparatus was organized in a way that facilitated continuous learning, about the enemy and about the operational environment. In this worldview, operations would feed intelligence and intelligence would feed operations, leading Flynn et al. to argue that today, "intelligence *is* operations."[10] To conjoin operations and intelligence into a seamless whole to facilitate knowledge development, the US military constructed a novel targeting methodology, known in military circles simply by its acronym, F3EAD[11] (find-fix-finish-exploit-assess-disseminate), pronounced FEED.[12]

Specifically designed to fuse intelligence and operations for the purpose of turning "hunting high-value targets into a high art,"[13] F3EAD would provide the logics and a framework to draw together and operationalize a number of micropractices and techniques. Gradually configured into a larger apparatus, F3EAD would enable "track 'em and whack 'em" on a massive scale, dispersed and distributed across different spaces at different speeds and intensities. Stanley McChrystal described the F3EAD methodology as such: "analysts who *found* the enemy, drone operators who *fixed* the target; combat teams who *finished* the target by capturing or killing him; specialists who *exploited* the intelligence the raid yielded, such as cell phones, maps, and detainees; and the intelligence analysts who *turned* this raw information into usable knowledge.[14]

The practice and processes of military targeting had now become the centerpiece in war-fighting efforts against the supposedly elusive networked enemy. Through recurring targeting cycles that would not only focus on the operational finishing aspect (kill/capture of targets) but also, it was believed, gain new data to set new analysis in motion, the military had configured a new methodology to learn about the enemy and the operational environment to facilitate ongoing (re)enactments between us and them and to unearth new targets at a rapid pace.

This signaled an important shift—not only in operational thinking but, more fundamentally, in martial epistemology. Instead of planning operations according to information that was carefully produced, processed, and analyzed, selecting targets according to clearly delineated effects, military forces now had to fight for rather than solely with information. Learning, rather than destruction, had now become a key operational task.

F3EAD would be designed to stimulate reactions through constant perturbations and feedback loops in order to learn about the enemy and the operational environment. As military scientist Mitch Ferry explains,

> Targeting is used now not merely to neutralise threats or prepare the operational environment, but to learn about the adversary. Whereas learning had previously been a means to an end (to enable a strike), it is now *an end in itself*. What targeting achieves has changed, bringing opportunities and threats related to the dual outcomes. Targeting in a complex environment *invites a tension between learning in order to affect and affecting in order to learn*.[15]

Although hunting down "known" targets—individuals tied to specific identifications with name and face, such as Osama bin-Laden, Ayman al-Zawahiri, Abu Musab al-Zarqawi, and Saddam Hussein, and the individuals on the infamous "Iraq Most Wanted Identification Playing Cards"—proved challenging, the main operational challenge, according to the US military, was those individuals that remained anonymous but were assumed to be out there, lurking in the dark. Because the enemy was seemingly operating below the detection thresholds of the martial sensorium, one needed to, the logic goes, force the enemy to emerge from its potential and materialize itself to be detectable by the targeting configuration and its epistemic operations.

Michael Hayden, recalling a conversation over dessert with the former commander of US Special Operations Command (SOCOM), Charlie Holland, gets to the crux of the challenge the F3EAD methodology was designed to meet. When confronted with SOCOM's need for more "actionable intelligence," Michael Hayden replied, "Charlie, let me give you another way of thinking about this. You give me a little action, and I'll give you a lot more intelligence." He clarified: "In other words, we needed operational moves to poke at the enemy, make him move and communicate, so we could learn more about him." And as simple as that, they "more and more settled into that pattern [where] operations [would be] designed to generate information."[16]

Settling into this pattern not only generated information but created what John Nagle described as "an almost industrial-scale counterterrorism killing machine,"[17] which had horrendous consequences for the countries, cities, villages, families, and individuals caught up in its web of operations. And like in Vietnam, this proved to be a disastrous failure.

The Learning Cycle

F3EAD was incorporated into the draft version of the US Army targeting doctrine, ATP 3-60, in mid-2008 and later into the 2010 version of the document, thus making it a recognized and approved methodology after years of operational experimentation. ATP 3-60 describes F3EAD as a methodology and a technique that "works at all levels for leaders to understand their operational environment and visualize the effects they want to achieve ... especially well suited and ... the primary means for engaging high-value individuals (HVI)." But, importantly, and along the lines of the previous chapters, the doctrine argues, "[the] focus of targeting is not just to identify an individual who is a leader in the network [but to identify the critical node(s)] whose removal will cause the most damage to the network." The ultimate success, the doctrine continues, "is to remove sufficient critical nodes simultaneously—or nearly so—such that the network cannot automatically reroute linkages, but suffers catastrophic failure."[18] This is an important point cognizant of targeting's relational structural assumptions and logics about an operational strategy of *delinking*.

David Kilcullen, one of the chief architects of the US's counterinsurgency (COIN) and advisor to General David Petraeus, argued for "a global Phoenix program" to defeat what he saw as a global jihad network, which "comprises multifarious, intricately ramified web of dependencies."[19] "As the organic systems model of insurgency shows," Kilcullen goes on to argue, "disrupting this network demands that we target the links (the web of dependencies itself) and the energy flows (inputs and outputs that pass between actors in the *jihad*) as the primary method of disrupting the network."[20] What this concept of delinking means is that the war effort should be aimed at the infrastructure of the enemy, not the leaders or specific nodes, thus resembling the logic behind the Phoenix program and its focus on the VCI. This time however, it would be global in scope.

Firmly placing individuals and their social relations and links, rather than the population, at the center of the analytical and operational challenge, F3EAD highlights the difference between COIN and CT. While COIN produces the population, like the Hamlet Evaluation System (HES), to focus the efforts on governing the population,[21] or at least winning "hearts and minds" through security, development, and strategic communication strategies, CT changes the focus from the military aphorism of "governing oneself to victory" to that of "killing one's way to victory." This does not mean that the population did not matter to F3EAD, but the population in question here served as the background feature, or "noise," through which one had to filter out the bad from the good. One had to know the population to sieve out targets and produce the enemy, but it was not about producing the population as a technically abstracted governing object itself.

Thus, at the conceptual and technical plane, both COIN and CT share the important commonality of systematically identifying, sorting, and classifying the population into objects of action. On the tactical and operational level, COIN focuses on highly refined kinetic and nonkinetic targeting efforts designed to "identify and separate the 'irreconcilables' from the 'reconcilables'"[22] in order to pinpoint security-development interventions. CT, on the other hand, stresses the need to find, fix, and kill or capture individuals among the population, to disrupt, degrade, dismantle, and defeat (DDDD) enemy networks. CT thus poses different epistemological challenges for the martial apparatus.

The traditional intelligence and targeting cycles that were established and used during the Cold War, such as order of battle (OB) and the "enemy as a system," were largely linear in the sense that they proceeded from a requirement or an established fact or knowledge, to collection of data, to analysis, and then to a finished product—an object that was identifiable or known (such as those within the Bombing Encyclopedia) and that could be monitored or acted upon. F3EAD, or "network-based targeting," as it is also known[23], on the other hand, was a major shift in reporting and assessing the identifiable (known knowns) to finding the hidden and often unknown (known unknowns and unknown knowns) through analytical search and discovery cycles.

Unlike the traditional targeting methods and processes, such as the US Army's D3A (decide-detect-deliver-assess) or the US Air Force's F2T2EA (find-fix-track-target-engage-assess), focusing primarily on "finishing," the main effort of F3EAD is the "exploit-analyze-disseminate" aspect that, it is believed, offers insights into enemy networks and generates new lines of operations—to find, fix, and finish the next target. Importantly, as Mitch Ferry

FEEDing the Network: Hunting for Signatures 133

contends, F3EAD "provides a method for dealing with high tempo operations against low-signature targets while focusing not only on the execution of the current target, but *the generation of new targets.*"[24] In other words, the major shift in targeting that was introduced with F3EAD was that it relied on an emergent understanding of the enemy and that it was specifically designed to perpetuate the targeting cycle and extend it to the search for unknowns through direct actions.

To transform manhunting into an abstract technical exercise, F3EAD uses multiple feedback loops to detect, stimulate, and uncover knowledge about the target system through recurrent targeting cycles. These feedback loops are both fast, in the immediate exploitation of information discovered, and slow, in the more thorough analysis phase, enabling so-called follow-on targeting cycles from the finish phase of a target. In F3EAD every potential target (of killing or capture) is also a potential source of intelligence—targets are produced and selected not only for the perceived threat they pose or for their relational systemic functions but also because of their capacity for intelligence exploitation for knowledge about the target system and thus generation of new targets. As such, F3EAD constituted a major shift in intelligence and operations from reporting on the identifiable to believing that the hidden could be found through cyclical learning processes.

In seeing feedback and circulation of data within the configuration as the essence of learning, "designed to enable organizational learning about the adversary and the environment,"[25] F3EAD modified the understanding and patterns of action of the martial apparatus based on past experience, but also on new data. This learning was, of course, situated in the dynamics of interaction between us and them, taking place in particular locations, situations, and contexts and from embedded positions and based on relations and interactions with the environment.[26] It was situated in the tacit knowledge of military operators and analysts, the doctrinized knowledge of the military institutions, military imaginaries of the network, but above all, anchored in the targeting configurations and their epistemic operations through which this learning was actualized.

The emphasis on learning would, in the minds of the US military, serve two interlinked purposes: first, to ensure the gathering of "the maximum information possible on the enemy network" which, second, would enable "a rapid operations tempo against the enemy."[27] The speed of multiple simultaneous operations against nodes in the network, the assumption goes, would sever links and deny the network the ability to adapt, forcing the enemy to focus on staying alive rather than focusing on sustaining their own operations. In order to do so, however, one had to learn about the network in real time. Because F3EAD assumed an emergent understanding of the enemy through operations, firmly connecting learning and destruction, it relied on several epistemic and technological practices, processes, and methods to slowly learn about and produce the enemy network.

With a novel war-fighting theory established and translated into doctrine and concepts of operations through the F3EAD framework, the next crucial task was invention, adaptation, and experimentation with specialized technical tools and techniques to reduce "anonymity on the battlefield, penetrate complex networks, and differentiate between friend and foe."[28] The production of the enemy thus required novel configurations of humans and technologies to support the techniques and practices needed. Some battle-hardened practices stemming from US operations in Vietnam, such as social network analysis (SNA) and pattern-of-life (PoL) analysis, were dusted off and updated with novel ways of processing various forms of heterogenous data. Coupled with novel data sources (such as the massive amount of metadata collected from communication traffic; human intelligence; biographic, biometric, and forensic data; a variety of open-source intelligence; and, not least, the massive amount of data from live video drone feeds) the US martial community was now famously "swimming in sensors and drowning in data."[29]

The problem was not so much coping with this data deluge, they argued, but being able to ingest, fuse, and process it together to produce a figure of the enemy and distill certain targets. Utilizing a variety of big data tools and techniques and creating machine-to-machine interfaces to automate as much of this processing as possible became the solution. This solution was the product of a specific historical configuration of a "manufacturing" process that relied on advanced algorithmic modeling and machine learning techniques for analysis. Reflecting on these violent transformations some ten years into the so-called global war on terror, Lieutenant Colonel John Nagel, stated,

> Counterinsurgency doctrine believes in killing people, it just believes in killing the right people. And what's happened over the past five years is we've gotten far, far better at *correlating human intelligence and signals intelligence to paint a very tight, coherent picture of who the enemy is and where the enemy hangs his hat*. And we've gotten better at using precision firepower to give those people very, very bad days. And I really think that this is redefining what counterinsurgency means in the 21st century.[30]

From Social to Threat Networks

The go-to tool for painting a very tight, coherent picture of who the enemy is and where the enemy hangs his hat was, like we saw in the previous chapter, social network analysis.[31] Like the systems analysis conducted previously, network analysis was not merely a metaphor for rendering complex patterns intelligible but became the key conceptual framework and knowledge practice for analyzing presumed enemy networks and identifying functional roles, organizational positions, and influential individuals.

Mareike de Goede argues that the network functions as a key security technology that depicts both the danger and the solution to the danger, rendering "the world actionable and amendable to intervention."[32] As such, she sees the network as both a mode of security knowledge and a risk technology in so far as it "classifies danger and ascribes riskiness to nodes and relations, with the aim of rendering future danger actionable in the present."[33] SNA, de Goede contends "aim[s] to transform the network into an actionable technique . . . seeking to generate targets in advance of attack, crime, or violence."[34] While de Goede links SNA and the network more generally to anticipatory and preemptive security regimes, military targeting, as this chapter will show, is less concerned with preempting and anticipating futures than it is with using the network as a tool for producing and operationalizing the enemy model, or, creating what is known in the intelligence community as "actionable intelligence".[35]

Based on theories about organizational structures, leadership, and centrality of nodes in networks, SNA is claimed to be able to identify those individuals (e.g., terrorists) that are most crucial to the operation of the network, be it financiers, bomb makers, couriers, high-level commanders, or leaders. What is created through SNA are static identities of nodes and their ties to each other, not the relations themselves and how these might change over time. SNA is not interested in the *why* of the network, only the *how* of intervening upon key nodes believed to maximize the disruptive effect. De Goede points out that this type of research has the apparent ability to "generate endless new investigative leads in security practice,"[36] naturally leading to a never-ending cycle of new targets being generated.

While this is a valid critique, it is not that target analysts are not aware of these limitations of SNA and traditional link analysis and contact chaining. As former CIA target analyst Nada Bakos put it in her memoirs, "I discovered early on that if I went too many levels deep with link analysis, I could've tied a bombing back to my dog."[37] Echoing de Goede's critique that "networks cannot exist without narratives that accord meaning to the relational ties that are identified and generated through its methods,"[38] Nada Bakos argues that "link analysis in and of itself is a reasonable tool for trying to define a landscape, but the analysis behind the chart is what really matters."[39] However, as this chapter will highlight later on, the CIA and the

US military differed in interpreting the data coming out of these systems, with the SOCOM favoring speed of action over deliberation and narration.

During the war on terror, SNA was gradually replaced by its updated version: threat network analysis (TNA). This shift was undertaken largely to emphasize the need to look not just at individuals. As former intelligence officer Robert Clark argues in his book, *Intelligence Analysis: A Target-Centric Approach*, this was important in order to see how all things interact, not just people. In TNA, nodes can be everything: "associated organizations, weapons, locations, and the means for conducting terrorist operations (vehicles, types of explosives)."[40] It works in the same ways as SNA, but with the added elements of nonhuman things, which presumably gives the network more detail.

Thus, as Clarke explains, the work consisted of developing "a detailed understanding of how a threat network functions by identifying its constituent elements, learning how its internal processes work to carry out operations, and seeing how all of the network components interact."[41] The structure of the network, with an emphasis on core-periphery, centrality, diameter, and hubs and their density, determines the weaknesses of the network and the approach needed to make the network collapse. Thus, Clark asserts, TNA is a better concept for analysis, because it links the operational environment, nodes, edges, and links, rather than more narrowly focusing on the nodes that make up the network.

Without drawing the analogy too far, it is worth mentioning that the move toward TNA is in many ways similar to the critique offered by actor-network theory (ANT) scholars against more traditional social network analysis in emphasizing the need to take nonhuman entities more seriously and focus on the relations between humans and nonhumans rather than the entities themselves. In TNA, as in ANT, one could then argue that there are no actors, in the sense that they do not have an essence or properties independent from their relations. They become what they are through their networked relations. In TNA, it is the relational links that determine what they can become, but at the same time there can be no structures that are independent from the elements that they connect. This gives rise to the rather peculiar and complex notion that nothing can be calculated about networks without considering every node, and little can be calculated about nodes without considering the network in its entirety. Or in the more cryptic words of Bruno Latour, "An actor is nothing but a network, except that a network is nothing but actors."[42]

The point here is not to argue that the US military has co-opted ANT to conceptualize and produce its enemies, as in Eyal Weizman's famous study on how certain elements within the Israeli military have drawn on Foucault, Deleuze, and other critical scholars;[43] they most certainly have not. Rather, it is to argue that the US martial community has largely drawn the same conclusion: that the anthropocentric SNA does not capture how these networks hang together or what capacities for action arise out of particular relations. The implications of TNA's way of translating and capturing the enemy is that it creates a whole different network, one that captures not only individuals and their links to other individuals but also their lifeworlds and ecosystems, reconstructing the operational environment and filling it with things *and* humans. Somewhat simplified, TNA combines the imaginaries and epistemologies of the enemy as a system with that of social networks.

Rather than just focusing on the centrality of certain individuals, TNA thus expands the enemy model and the operational environment and also the scope of both what is of interest to targeting and what a target can be. By calculating, classifying, ordering, and visualizing everything in a tangible format as material network connections, TNA buries deep into socio-material structures of modern life, expanding the notion of battlespace and creating novel military space-times. Although nodes were the tangible objects through which the military could direct action, it is not the targets in themselves, whether individuals or fixed sites, that are the most important, but their links and relations and especially what capacities of action arise from these—hence, the focus on severing the links rather than taking out specific targets. This is not to deny the visible effects of targeting, which are almost always the destruction or killing/capturing of a specific target, but if we are to understand contemporary targeting and warfare more generally, we must understand how the enemy is modeled rather than focus just on the ways in which individuals are found, tracked, and killed.

The argument is that network analysis is not only a mathematical network modeling practice that reifies the network as a measurable and calculative entity, but it also, as De Goede argues, promises the ability to map whole networks, produce the insights needed for targeting, and actualize the nodes and links it portrays.[44] This promise is largely backed up by producing clarity and simplicity through the abstraction of reality and visualizing the model in a manner that is accessible to action, often through very simple causal logics—take away this and that node and this link will be severed. However, the abstract models that are built obscure the uncertainties of epistemic operations that these networked understandings rest on. As a result, the wrong people were constantly made up.

This was, however, not only because of the network logics endlessly generating new leads. It was largely because F3EAD worked from an emergent understanding of the enemy in which the logic of perturbating the environment in order to learn became commonsensical. The result was that the US military extended its epistemic and military operations indefinitely and everywhere, creating an illusion of the systems working toward victory. How this came about is important to describe in more detail.

The "Find"

Although the enemy is mapped as a threat network consisting of things and individuals, so-called high-value individuals (HVIs) are the focal point of most F3EAD operations. These are individuals of interest who the military deem important enough to be identified, surveilled, tracked, and finally, "influenced" by either information operations or fires.[45] Thus, HVIs only become targets for kill/capture operations after an evaluation and are often then placed on a target list. Such lists, like the infamous disposition matrix and the Joint Prioritized Effects List (JPEL)[46], classify, rank, and prioritize targets according to a method called CARVER (criticality, accessibility, recuperability, vulnerability, effect, and recognizability) to assess the effects of killing, capturing, or otherwise influencing that person on the overall threat network. Dating back to WWII, CARVER is simply a method for weighing the value of a target's importance, or criticality, to the insurgent cell.[47] It functions as a martial mnemonic that aids analysts in their prioritization of which targets to influence. However, before targets reach this stage, several steps are carried out.

Although the word *find* in F3EAD brings to mind a game of hide-and-seek, hunting or foraging, targets are not just found, but produced. Most of the targets produced through targeting are people without a clear biographical or biometrical identification; they are individuals that are produced as targets either from being relationally linked to other suspects or because they display certain signatures that are predetermined to portray animosity. According to Flynn and colleagues, FIND "describes the process through which targets and their elements are detected, modelled, and prepared," including "target system modelling (identifying detectable and targetable system elements), pattern of life development, target nomination (detailed request to target execution authority—based on a 'target pack' including intelligence and analysis) and preliminary target location (used to trigger collection assets to locate and identify target elements)."[48]

The find step in F3EAD thus makes up an HVI and identifies and maps its network and, finally, produces a targeting folder that includes what is called a "baseball card"—a concise summary of the supposedly key information on the HVI that can then be passed downstream as an input in the F3EAD algorithm and inserted into the "fix" element.[49] These baseball cards consist of biographic and biometrical information, a map of the area in which the HVI is presumed to be, a picture of the HVI, their personal history, pattern of life (the where, when, who, what, and how), a mobile phone number, known associates and position within a network, and other information deemed necessary (see figures 5.1[50] and 5.2[51])—filed entries that are very similar to the ones compiled in the Bombing Encyclopedia and the Phoenix program folders, technically abstracted, stylized, and fixed. However, these baseball cards do not take the form of a finished piece of intelligence that is ready to be acted upon but take the form of a lead through which the fixing step of the cycle can start.

To find an enemy whose "primary strength is denying friendly forces access to a target,"[52] hiding "in plain sight," "among the population," or "in the midst of civilian clutter,"[53] the martial apparatus had to reconfigure their operations/intelligence machinery to incorporate previously disparate "INTs" and collection methods, such as SIGINT (signals intelligence), HUMINT (human intelligence), GEOINT (geospatial intelligence), OSINT (open-source intelligence), and their various sensors. This largely involved an exploration of all sources at their disposal, including traditional HUMINT such as tips and patrols, imagery from various optical sensors, radars, and airborne reconnaissance missions, interrogations of captured enemy combatants, signals intelligence and massive interception of telecommunications, both content and metadata, and open-source intelligence from publicly available information such as social media platforms.

As the targeting doctrine explains, "Signals intelligence for example, can locate a target but may not be able to discern who it is. An airborne sensor with full motion video can track but not necessarily identify the target. Human intelligence can provide intent but may not be able to fix a target to an accurate location."[54] Used together, however, it is believed that this targeting assemblage can "provide a start point into the enemy network that can be exploited through persistent and patient observation."[55] The key to the F3EAD methodology was, according to doctrine, "massed, persistent reconnaissance, or surveillance cued to a powerful and decentralized all source intelligence apparatus to find a HVI."[56]

The 3-60 targeting doctrine highlights two important techniques in the find step of the F3EAD process: *nodal analysis* and *life pattern analysis*. Nodal, or link, analysis, also known as call chaining or contact chaining, depending on whether it is a phone or an individual that is chained, is simply the technique through which one connects individual elements to each other in order to populate the presumed network structure and reveal its power

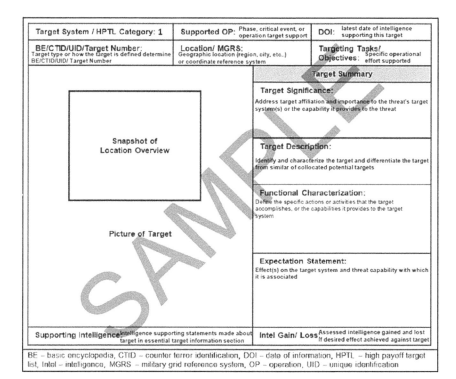

Figure 5.1

Baseball card 1. *Source*: US Army, Army Techniques Publication (ATP) 3-60 Targeting (May 2015): D-10.1.

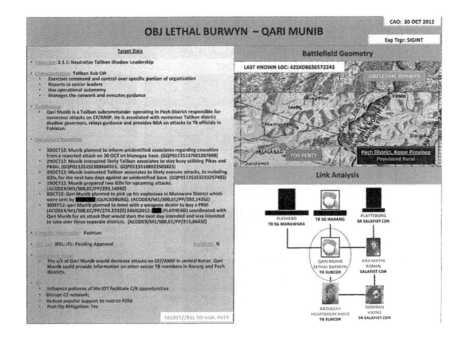

Figure 5.2
Baseball card 2. *Source*: Josh Begley, "The Drone Papers: A Visual Glossary," *The Intercept*, 2015, 2.

relations. This requires what is known as a starting point, a selector, or a seed, through which the analysis can begin.

A selector in SIGINT language is a signal identifier such as a telephone number, the unique IMEI (international mobile equipment identity) number for all mobile devices, IMSI (international mobile subscriber identity) numbers, MSISDN (mobile station international subscriber directory number), e-mail address, or instant messaging ID. When selectors are identified as suspicious, or have been selected as reasonable, articulable suspicion (RAS)—the legal synonym for probable cause—they become seeds.[57] In addition, seeds may just be individuals identified as suspicious through HUMINT or interrogations. Once a seed is known, one can start by shallowly looking at which phone numbers (or other kinds of identifiers) or individuals are in contact with each other.

This analytical process, according to a UK Government Communications Head Quarters (GCHQ) handbook on data mining, is the start of a "painstaking process of assembling information about a terrorist cell or network."[58] The National Security Agency (NSA) usually entered the seed into contact chaining systems (such as MAINWAY) to retrieve all the numbers that had been in contact with the seed. Linking these seeds to other phones and people, it was believed, would cause a network of individuals to slowly emerge.

The intelligence community was wary that too many "hops" from the seed would lead to large and complex networks that were too unwieldy to investigate and usually limited themselves to two hops from the seed and a maximum of three.[59] These chains or links would then be visualized. Through commercially available software programs such as IBM Analyst Notebook or Palantir Gotham and other contact chaining tools created for and by the martial apparatus such as Renoir, network diagrams or tree-like structures that grow as more and more data are added, would produce abstracted phantasmatic models of the enemy.[60]

Contact chaining was thus a tool that was reliant on a known, stripped of context and narrative, whose social and material relations, it was believed, would provide insights into the enemy's networks, leading to further starting points that could be observed and tracked.[61] It is a method that operates exclusively on quantitative data, and in particular on metadata, for example the number of phone calls made to and from an existing known. According to Chamayou, the practice of contact chaining creates a target that is "conceived fundamentally as a reticular individuality."[62] It is thus incapable of recognizing qualitative distinctions and differences among the relationships depicted in its network diagrams. The result was that individuals were forced together

FEEDing the Network: Hunting for Signatures 143

on the basis of their quantitative relationships—relationships that would multiply tremendously in just a few hops. If we suppose, for instance, that one phone has been in contact with 100 other phones and that each of these 100 phones has been in contact with another 100 phones (two "hops"), a quick calculation takes us up to over 10,000 phones, each with a person at the other end. If we add another "hop", the network of phones quickly multiplies to 1,000,000. This was the network that was created out of a single phone. While linking individual nodes appeared clean and seductive to decision-makers, in reality, it disguised as much as it revealed, creating a fantastical model of the enemy that would constantly make up the wrong people.

The Life Patterns of Nodes

Another technique for disclosing enemy networks favored by the martial apparatus is life pattern analysis (PoL)—the situated data generated by observing daily human behavior. According to US targeting doctrine, it is the technique through which the US military "is connecting the relationships between places and people by tracking their patterns of life."[63] This tracking technique, like contact chaining, is predicated upon a known seed that is then persistently surveilled by various means between locations and other individuals. Once leads are identified, various fixing tools, from full-motion video to algorithmic modes of observation and automated metadata analysis, are put into play. Unlike the time-intensive pattern-of-life analysis that was conducted during the Vietnam War to fix suspects, PoL today utilizes complex airborne surveillance sensors linked to various software tools working on large data sets that have taken over the job of direct surveillance—to enable what is known in the military as "The Unblinking Eye."

According to Flynn, "the Unblinking Eye provides an opportunity to learn about the network in action and how it operates. . . . The purpose of this long dwell airborne stakeout is to apply multisensor observation 24/7 to achieve a greater understanding of how the enemy's network operates by building a pattern of life analysis."[64] These are designed and engineered to track individuals in physical or virtual locations or observe geographical locations in real time on a 24/7 basis to identify and establish patterns of life. What Flynn et al. calls nodal analysis,

> is spatially connecting relationships between places and people by tracking their patterns of life. . . . Connections between those sites and persons to the target are built, and nodes in the enemy's low-contrast network emerge . . . revealing his physical infrastructure for things such

as funding, meetings, headquarters, media outlets, and weapons supply points. As a result, the network becomes more visible and vulnerable, thus negating the enemy's asymmetric advantage of denying a target.[65]

PoL thus rests on the belief that gaining deeper insights into the daily life of people or areas will reveal the structure of threat networks. One can thus say that while contact chaining is a tool for linking individuals and their phones to each other, thereby building a social network, PoL supposedly makes up a more comprehensive threat network to include all entities, such as infrastructure and geographical location, by tracking certain behaviors of humans and things. While PoL is also extensively used to fix an individual so that a strike or raid can be planned, a "finish" operation only happens when the target is determined to be ripe and all information valuable to the F3EAD process has been exhausted from it.

Fixing an individual using PoL analysis requires the establishment of a baseball card (classifications, attributes, and relations) that works to foreground the fix and the finish. The abstracted features of this entity, we are told, can determine the patterns of space and time in the course of its normal behavior against a contextualized environment, or background. That is, an individual's or location's pattern of life is thus a logical and technically abstract representation of entities moving within spatial regions over time in which the frequency of the spatiotemporal patterns are defined as the logical sense of normalcy. In a more data-scientific way, what is tracked are the spatiotemporal tracks, or the geometric patterns of spatial locations and temporal patterns of timestamps and their duration.[66] When these patterns repeat themselves over time (similar temporal durations and spatial patterns of movement), normal patterns of behavior are established, and the target can be fixed, planned, and made actionable.[67]

The difference between activity and behavior is important here. Activity is usually associated with what the intelligence community (IC) refers to as a "sensor event," which can be a one-off occurrence, like the triggering of an alert system, as with the Igloo White from the previous chapter. Behaviors, on the other hand, are representations of an entity's activities extracted from available data and observations and are usually something that only occurs over time.[68] That is, behavior is understood as activity with purpose or meaning, and it is believed that by monitoring activity over time one can understand behavioral patterns and, from there, infer the identity and intent, good or bad, of that entity. The target that emerges through PoL is thus a highly convoluted relational type of entity that exists only as a spatiotemporal pattern, or a state of affairs drawn from the datafied contextual background. Here is Joseph Pugliese vividly describing the martial fabrications of PoL:

> The military term "pattern of life" is inscribed with two intertwined systems of scientific conceptuality: algorithmic and biological. The human subject detected by drone's surveillance cameras is, in the first scientific schema, transmuted algorithmically into a patterned sequence of numerals: the digital code of ones and zeros. Converted into digital data coded as a "pattern of life", the targeted human subject is reduced to an anonymous simulacrum that flickers across the screen and that can effectively be liquidated into a "pattern of death" with the swivel of a joystick. Viewed through the scientific gaze of clinical biology, "pattern of life" connects the drone's scanning technologies to the discourse of an instrumentalist science, its constitutive gaze of objectifying detachment and its production of exterminatory violence. Patterns of life are what are discovered and analysed in the Petri dish of the laboratory.[69]

In essence, if a person's lifestyle tells the martial apparatus that there is a certain percentage chance of that person being a terrorist, or associated with terrorist activity, they have the right to kill or capture that person.

Making Up the Wrong People

Let me further illustrate this story of petri dishes with a chilling account from Afghanistan, drawn from the book *First Platoon* by investigative journalist Annie Jacobsen, detailing how pattern-of-analysis operations worked—or did not work.

> "I could see everything," Kevin says, referring to the aerostat's technology-enabled omniscience, sometimes called the God's-eye view. "The only way I didn't see something was if I wasn't looking at it." ... Working as a pattern-of-life expert at Sia Choy, Kevin spent hours a day, seven days a week watching all kinds of people go about their lives, with the goal of separating out the insurgents from the civilians.

The story is set in 2012 and follows Kevin, a mission director for the Persistent Ground Surveillance System (PGSS) team in the Zhari District in Afghanistan. His daily task was to utilize the various sensors and software tools of the PGSS to produce data and persistently observe vast areas and individuals of interest. After watching the particular rhythms of a man in a purple hat for weeks, Kevin and his team had determined that the man in question was placing improvised explosive devices, or IEDs, in the ground for the Taliban.

"We elevated him to 429 status through his actions." 429 status is what happens when a person of interest completes three "interactions with the ground." These are actions that allow for that individual to be moved out of civilian status and into insurgent status—to be targeted and killed legally according to army rules of engagement. "If I see him interacting with the ground, and then I see the pressure tank going in, or [if] I see the charge going in, and him stringing the lamp cord out to install his pressure plate or his battery connection." . . .

"That's one, two, and three for me," Kevin says.

The result of this persistent surveillance was thus a "three strikes out" logic, which moved the individual from suspect to kill/capture operations. But not before the PGSS team had produced a PowerPoint containing all the known data about this person, like the baseball cards (figures 5.1 and 5.2), and sent it to the army's S2 intelligence officer. After a quick review, this information was sent to the battle captain.

> "He takes that info," Kevin explains, "and he washes it through Palantir." . . . "The military application of Palantir is awesome," Kevin says. Palantir is capable of mining and aggregating data on individual people in a manner that would astonish almost anyone. . . . Once the PGSS team located who they thought was the actual person of interest, "we'd kind of do a self-check, to follow him." Meaning the initial hunt began with a computer, but it was now fact-checked by a human. . . . "The data cuts from Palantir are like a bread-crumb trail for me to go down. At the same time, if I see something, then that's me generating a report. And that becomes data in Palantir."

This step-by-step process—entangling data seeds from Palantir software with activity determined as suspicious by so-called persistent surveillance— is somewhat reminiscent of Henry Nash's experiences (chapter 3) of being a target analyst in the Air Target Division during the Cold War, obscuring the relationship between data and outcome, cause and effect, in which individuals were prevented from understanding the larger picture. By compiling a PowerPoint, or baseball cards of individuals, enclosing highly abstracted and standardized data, operators like Kevin had limited knowledge on what these data truly represented—and, importantly, how this information would be used. One thing was certain, however; once a person was made up to be a "429 package," the "tracking and whacking" was activated.

> "If there is an asset available, if CAS," close air support like attack helicopters and drones "is in the vicinity, then it is time to take the

target out." If there's not air support available, then the person of interest remains marked for death in the system. "The moment there is a target of opportunity to take him out, I call it in. I don't have to go back through the approving process," Kevin says. "The 429 package stands. That's why it's called a Target of Opportunity. When you have the opportunity, you strike the target." You kill the man.

Making up, tracking, storing, "marking for death," and "whacking" individuals was the core business of the global manhunts activated by the so-called war on terror, and Kevin was part of these end-to-end operations, creating opportunities for targeting and targets of opportunity. One morning, a window of opportunity presented itself to Kevin by way of an automated imaging system:

> "We're about to kill the man in the purple hat." . . . Standing in the C2 shelter at Siah Choy, in front of the video screens, the colleague spoke, "We're getting ready to hit him now," he said. "CAS is on the way." . . . "That isn't him," Kevin said. "That is absolutely *not* him." Kevin was certain of this. . . . Kevin ran out of the operations center, across the outpost and into the tactical operations center. "I told the S2 they had to call off the air strike. It's not him," Kevin told the battle captain . . . S2 [eventually] called off the air strike. "I doubted the computer. I was right." How was the farmer on the tractor misrecognized as the cell leader in the purple hat in the first place? "It was his hat," Kevin explains. "There's a window of time, around dawn, as the sun comes up," he explains, where colors are "read differently" by the imaging system than how it sees them during the day. In this window of time, the farmer's hat was misidentified as purple, setting off a series of linkages that were based on information that was erroneous to begin with.

This story illustrates much of what has been analyzed so far, showing the messy ways in which people are made up and potentially killed—from small bits and pieces of data, "like a bread-crumb trail," to 24/7 PoL surveillance, "info" "washed through" commercial data analysis programs, to the configurations of humans and machines feeding each other with data and confirming the validity of the targeting process. While this is a story that saved an innocent man's life, most were not that lucky. These targeting configurations, automating perception, memory, and knowledge production, constantly made up the wrong people.

"SIGINT is the Coin of the Realm"

When collaging and reading various martial documents, from military doctrines, handbooks, and technical reports to industry promotions and the cache of documents leaked by Edward Snowden and others, a vast and messy assemblage of different technologies and techniques and methods for producing, processing, and analyzing data in order to produce and operationalize the enemy comes to the fore. These are just some of them.

In the air, planes were outfitted with a growing family of SIGINT and COMINT sensors, with codenames such as NEBULA and WINDJAMMER, which swept up massive amounts of data (metadata, content, and geolocation data) and fed it back to ground stations, populating large searchable databases at the disposal of the US IC and its analysts. Sensors could also be remotely tasked by analysts and technicians in theater or even from the US homeland, coordinating with air crews using communications platforms such as mIRC, and send the data directly to targeting teams. On the ground, so-called low-level voice interception (LLVI) teams, often embedded with combat units, were monitoring radio traffic and offered location directions to the SOF hunting teams. In cyberspace and space, data was swept up by bulk collection sensors and systems that were directly tapping off cables and satellites and populating databases. Most of the targeting process were undertaken using various SIGINT and COMINT databases, cross-referenced with geolocation data, and supplemented with imagery and HUMINT data. Once this was complete, HVIs become HPTs, or high-priority targets, and added to the disposition matrix and the Joint Prioritized Effects List (JPEL), ready for kill/capture.

Although multi-INT was recognized as important, SIGINT was the "coin of the realm"[71] in the war on terror. A core component of this transformation was the concept of "persistent surveillance," which served the mantra of the then–NSA director General Keith Alexander to "collect it all". This led to a new intelligence posture: "Sniff it all, know it all, collect it all, process it all, exploit it all, share it all."[72] Persistent surveillance was in large part enabled by novel ISR (intelligence, surveillance, and reconnaissance) surveillance technologies such as airborne sensors, but increasingly, the NSA realized that the digital age presented itself as the "golden age of SIGINT."[73] The digital age allowed the NSA to augment their decades-old collection platforms that were mainly designed for passive interceptions of communications or activity with more active operations that allowed for mass collection of data and information that resided in sensitive and private systems.

An interesting difference between this heterogenous "collect it all" haystack and the Bombing Encyclopedia, the HES, and the Phoenix databases is that what kinds of data are of interest and relevant is not predefined. While the latter databases were produced through carefully standardized formats such as the Consolidated Target Intelligence File (CTIF) and the Hamlet Evaluation Worksheet (HEW), which largely knew what type of data they were interested in, today's databases are built through heterogenous data from multiple sources and source types which are only fused together *after* the collection stage, at the processing and analysis phase. The various NSA documents speak about ingestion, rather than collection, of data, meaning that it is intended to scoop up any kind of data, not necessarily produced for the specific purpose of targeting, like the trained human sensors of the HEW and the BE and their standardized formats.

What they were particularly interested in was so-called metadata, or data about data. It was believed that this could discern links and chains among objects of interest, but this required an aggregation of multiple different intelligence systems that existed in the US war machinery. The core idea behind persistent surveillance was to fuse different collection systems and data to detect, collect, disseminate, and characterize activity[74] and search this so-called multi-INT aggregated data to find associations between people, events, and activities, with the objective of uncovering patterns of behavior to drive more focused collection and investigation.

Analysis in the F3EAD cycle thus resulted in actionable intelligence that was deployed immediately to target an enemy individual or activity but that was also utilized as part of a more in-depth assessment of how a specific network was organized in a specific place. Importantly, the find step of the methodology are used not only to identify specific actionable targets but, notably, to "uncover the nature, functions, structures, and numbers of networks" within the operational environment, which allegedly allows commanders to gain understanding about the operational environment and to also visualize the threat network.[75]

While the classification technologies of the Cold War demanded the stability of categories within certain rule-governed manipulable parameters, the categories of the "it all" big data machinery is one of constant modularity, defining and redefining the categories according to context and the data logics through which they are actualized. So instead of homogenizing and routinizing the work of individuals collecting and analyzing data, the collect-it-all technologies were designed, besides from acting outside of human possibilities in speed and workloads, to allow for heterogeneity of data *and* cognitive drifting, experimentally combining and interpreting information and connect it with meaning. This created fuzzy boundaries between humans and machines, distributed agency in novel ways, and transformed the contexts and conditions under which human cognition operates within the system, altering its role in the production of meaning.

The functional fusion of operations and intelligence under F3EAD led to a situation that effectively created a self-sustaining logic based on the idea that if one collects, stores, and process everything all the time, the enemy will be made known. As Alexander put it, "'Rather than looking for a single needle in a haystack . . . collect the whole haystack. Collect it all,' he declared. 'Tag it, store it, and whenever you want, go searching in it."[76] Far from being simply a technical objective construct, the imaginaries and logics that went into these targeting systems made this process a martial technology. Jutta Weber argues that this type of othering through data mining has a built-in techno-rationality that fosters the endless production of possible future targets in a data-driven killing apparatus that consolidates and advances a possibilistic, preemptive culture of techno-security.[77]

The infamous disposition matrix, Weber argues, "can be understood as a constantly evolving database, which includes not only biographies but also conclusions drawn from data analysis, primarily on the basis of metadata" in which computational processing is intertwined with human decision-making in the production of targets.[78] Thus, these systems are about not merely a managerial logic, streamlining and institutionalizing cognitive work, but, rather, a product of its own networked model of the world, shaping the relationship between instruments, machines, and the environment in unforeseen ways. These various searchable databases and the automation of collection and data mining became so essential to the F3EAD process and the war effort in general that they came to, in Graham Harwood's words, "transform the war into a relational query, and the relational query into war."[79]

However, it is important, as Grégoire Chamayou argues, not to confuse the "collect it all" mantra with the language of total surveillance, Big Brother, panopticon, and the all-seeing, all-knowing power of the NSA that surfaced in the wake of the Snowden revelations. While not only being empirically wrong, this is also the wrong conclusion, Chamayou insists; they have "neither the capacity nor the desire to *actively monitor the whole world*." He explains, "It matters less to the NSA to actively monitor the entire world than to endow itself with the powers to target *anyone*—or, rather, *whomever it wishes*. The surveillance targets will be defined according to the priorities of the day; that is to say, according to the needs of the 'programme', this time understood as the totality of 'actions that it aims to accomplish.'"[80]

Chamayou's argument is astute and can be backed up by the empirics this book has already shown. If we substitute Chamayou's notion of "programme" with targeting, it is clear that the imaginaries, epistemologies, and data logics that targeting rests on do not have monitoring of the whole world as their aim but, rather, aim to produce and operationalize the enemy and materialize it through targets. While this does not reside in the genealogies of surveillance, it does rest on producing a totality by ingesting and analyzing as much data as possible in order to produce parts. The result is, in other words, that in order to "target anyone—or whomever it wishes," the martial apparatus has to target everyone and everything, everywhere and all the time.

Hunting for Signatures

The new fluid figure of the enemy created further circular needs for advanced logical and mathematical reasoning that could be carried over and used against the modeled enemy. This, it was believed, would enable the martial machinery to uncover known unknowns and the unknown knowns—looking for predetermined suspicious behaviors in unknown areas or tracking known suspicious locations for unknown people. The identity inferred from monitoring behavior was usually cross-referenced against predefined assumptions about what constituted signatures of enmity. The target analyst's logic might run this way: "If person X is behaving like this, it is highly likely, based on my assumptions and experience, backed up by data analysis, that person X must be a terrorist, and then I can legally kill him/her." Thus, predefined signatures were used to define enmity, hostility, threat, or the intelligence capacity of those observed entities.

Derek Gregory, like Pugliese, links pattern-of-life analysis to the full-motion video (FMV) feeds provided by the watchful eyes of drones and analyzed by human onlookers.[81] But the FMV feeds, as shown, were only a small portion of the full range of intelligence collection sensors and analytical tools that were locked into targeting configurations. Advanced computational analytics and machine learning techniques were used to sort through an immense mass of heterogenous data—telecommunications, bank transactions, administrative files, social media, geolocation data, and so on—searching for signatures of terrorist behavior.

If these activities showed signs of being similar to already established signatures of enmity, one could be deemed a terrorist and subsequently be on the receiving end of so-called signature strikes—kills or captures based on predefined signatures of enmity and not on confirmed identities and biographical details about an individual, which was put in the category of "personality strikes."[82] As Thomas and Dougherty explain, "Previously, CT 'finishes' were usually personality-based, i.e., the target was identified by name as a person of interest. Increasingly, CT strikes may be based on certain threat "signature" activities such as behavioral patterns associated with terrorist operations."[83]

Most of the identities on targeting lists were not confirmed biographical identities but were certified through signature recognition based on patterns corresponding to known entities. These databases simply contained the output of these messy cognitive operations in which individuals were the result of algorithmic decomposition and reaggregation of biographies, traces, activities, locations, associations, correlations, and networks.[84] Although these lists are part of a formalized legitimizing and juridification process of turning signature strikes into personality strikes, in the belief that this could help to avoid the unintended consequences of using violence,[85] most individuals killed or captured were not on these lists. Rather, they were the "intended consequence" of rapid high-tempo F3EAD manhunts that were premised on turning operations into intelligence and intelligence into operations as fast as possible to disintegrate the network faster than it could regenerate.

This rested not only on the belief that the enemy was "out there" readily waiting to be found but also on the premise that there was something like a "terrorist signature" and that by combining contact chaining and PoL analysis with that of general searches for terrorist signatures, the enemy structure would soon become visible and be taken down. Besides the fantasy of there being such a thing as a "terrorist signature", the problem was that, unlike the ocean, the skies, or the jungles of Indochina, this new information space consisted of an immense mass of heterogenous data.

While signature strikes resemble conventional operations against predefined threat signatures that beam certain technical signatures—an incoming missile or aircraft (usually associated with radar systems), submarines, equipment (such as that from Igloo White), and facilities—individuals seldom gave away such technical signatures, even in the new information space. Presumed terrorist signatures of enmity were much more complex than the more simple technical signatures of, for instance, a missile, a submarine, or a truck driving down the Ho Chi Minh trail, and signature strikes were not triggered on the fly as a sudden response to an imminent threat. Nevertheless, as artist and critical media theorist Hito Steyrel argues, separating signals from noise is not simply about recognizing patterns but is largely about creating them in the first place.[86]

It was, as Pugliese points out, created in a petri dish of big-data sets, grown and experimented with—a sustained pattern-of-life analysis in which similarities to the predefined and preproduced warranted action. With relational databases functioning as containers of possibility, baseball-cards and data could be endlessly combined and recombined as new data came in and the network shifted its structures according to actions and reactions. The result of these messy epistemic operations was not only that people were wrongly made up but also that it largely neglected the sociopolitical environment and obscured the horrific effects of violence that could have made decision-makers think differently about the war.

The SKYNET Is the Limit

A great example of the hunt for signatures is the infamous NSA SKYNET program. It works like many other of the multitudes of NSA programs revealed by Edward Snowden: collecting metadata, storing it on searchable NSA cloud servers, and applying machine learning algorithms to process and extract relevant information to identify leads. SKYNET assumes that terrorists' behavior is significantly different from that of other people when they are piecing together and reaggregating people's daily lives into suspects.

According to a slide deck presentation on the program, SKYNET specializes in identifying unknown couriers, not militants, insurgents, or operational terrorists per se.[87] Although it is not clear why this is its focus, we can infer the reasons based on the F3EAD process in its entirety and the logics of delinking. Couriers function as connectors and facilitators between nodes and networks, bringing with them knowledge, finances, and information in and around clandestine networks that are below the visuality thresholds of the unblinking eye. Because of this supposed linking role, the military views couriers as vital for revealing the structure of the assumed enemy network as well as pointing them in the direction of specific nodes that are in hiding. Couriers are targets, but not for killing.

In essence, SKYNET is a tool for discerning patterns in bulk phone or Global System for Mobile Communications (GSM) metadata, including patterns of life, contact chaining, and travel patterns, and as the slides show, its machine learning algorithm uses eighty different properties for each phone user to numerically score and rate people on their likelihood of being a terrorist. The score is then measured to a predefined threshold value (e.g., if above = suspected terrorist, if below = normal). "Given a handful of courier *selectors*, can we find others that 'behave similarly' by analyzing GSM metadata?"[88] one of the slides asks rhetorically. The rest of the deck is used to answer an affirmative, "Yes, Sir!" The most interesting aspect of this program is not only the belief in the possibility of translating human behavior into a score but the way in which this technical abstraction works. Using a family of machine learning algorithms named Random Forest, which seem to be the algorithms of choice for much of the IC,[89] they start training their system on what the slides term "known Couriers + anchory selectors." Known couriers are in this rendering actual people that the martial apparatus "knows" are couriers and has distilled their habitual routines and behavioral traits into a set of knowns. Anchory selectors, on the other hand, seem to be specific preselected points in the training data used to optimize the training of the algorithms with so-called ground truths.

As a report in *Ars Technica* tells us, this is like training a Bayesian spam filter; you feed it known spam (couriers) and known nonspam (ordinary people). These "known couriers and selectors or signatures [are supposed] to serve as so-called ground truths through which the system learns to filter out suspects and produces a score based on Bayesian probabilities."[90] Spam filtering, Paul Kockelman tells us, works as a sieve—a "device that separates desired materials from undesired materials."[91] Algorithmic sieves are thus essential to information processing and sorting the world, producing patterns and predictability, but also meaning, through their cognitive operations. Kockelman argues that algorithmic sieves are important because of the ways in which ontologies, or ground truths, which in the case of SKYNET are known couriers plus anchory selectors, are embodied in and drive the interpretations by such algorithms.

Spam filtering, like the SKYNET, thus only works through the ground truth that general features, or known signatures, of spam are already known—and thus base their identification of individuals or spam on such statistical assumption (e.g., known couriers and anchory selectors). How ground truths are designed and engineered into the algorithmic sieves is thus important for how the systems interpret the information and connect it with meaning—a process that warrants a closer examination.

Matteo Pasquinelli tells us that machine learning techniques in general are based on formulas of error correction through the calculation of a statistical distribution of a pattern. At the core of these systems are three elements: the data, the algorithm, and the model. Basically, the learning algorithms have the purpose of computing a statistical model out of the training data. The learning here is based on what the model learns through the mapping of correlations, or patterns, between the input and the desired output.[92] SKYNET followed the same basic processes.

In the case of SKYNET the desired output is the known signatures, and the algorithms were, according to the slides, trained using a dataset consisting of 100,000 randomly selected people and seven known couriers. The algorithmic model was inscribed from six of these, as knowns or ground-truths, while the last known was the one the system was tasked to find.

After SKYNET had gone through extensive training on this dataset and the Random Forest model had been tinkered and experimented with to such an extent that an acceptable error rate had been reached, the slides claimed that their false alarm rate was as low as 0.008 percent.

Although this seems like a tiny percentage, we can do a simple equation on the possible effects. If we apply this to the population of, for instance, Pakistan, which is roughly around 230 million, we get around 18,000 people. If we say half of these own a mobile phone, we are down to 6,000 people who might be falsely suspected by the SKYNET. And this is, of course, only if the program is working at all in this way. In addition, what happens if we start treating these 6,000 as seeds and go one or even two hops away from the root? Then this dragnet starts to multiply in drastic fashion and the category of courier, militant, terrorist, or suspect becomes very flexible.

We do not know how known couriers were created, or the properties assigned to them, but the fact that Ahmad Zaidan, Al-Jazeera's longtime bureau chief in Islamabad, was given a top score when the system was trained on "+55M Pakistani selectors" points in the direction of travel patterns into predefined known terrorist spaces and would at least give one a good chance of a place on the SKYNET podium.

Regardless of labeling Zaidan as a member of Al-Qa'ida and assigning him a Terrorist Identities Datamart Environment (TIDE) person number, indicating inclusion in a central database of known or suspected terrorists, the final slide of the SKYNET presentation states that "preliminary results indicate that we're on the right track"—a result that "hopefully is indicative of the detector performing well 'in the wild,'"[93] a euphemism otherwise known as people's lives and habitats. Now that the training was over, SKYNET could be inserted into the wild and float inside the digital clouds, ingesting data and egesting scores, transforming the experimental correlations of Random Forests into a martial machinery of signature hunting.

What was created was in many ways an "algorithmic identity,"[94] a Deleuzian "dividual,"[95] most often unidentified by name or biographic features but identified through algorithmic modes of automated observation and collection and analysis of biometric features, activities, and behaviors correlated against predefined signatures of enmity. The problem, of course, is that SKYNET would overlook everything that was statistically different from the known couriers used to train the model. It would only be able to find those that were statistically similar to those knowns, resulting in a binary epistemology in which one ended up either as suspect or not. There was nothing in between. While Aradau and Blanke see SKYNET as an example of anomaly detection and the hunt for the unknown unknown, reading SKYNET through its use of "known couriers" places the program as part of the larger targeting assemblage hunting for known signatures.[96]

The main problem with SKYNET is not only its reliance on the Random Forest technique, or the strong beliefs in these machine learning methods, but the very notion that there is such a thing as a (or in this case seven) known courier/terrorist with attached behavioral signatures or that one can base a whole algorithmic system on. It does, however, enforce the idea behind looking at configurations, because in the case of SKYNET, it becomes clear that designing a system and inscribing and teaching a model based on these knowns is not only "bad science,"[97] and probably a waste of resources, but worse, extremely harmful when put into the wild—not only for those unlucky winners of the courier evaluation lottery, and for the first and second hops in the contact chain from this person, but also more broadly because of the ways in which these systems are designed to evolve and transform through their own operations.

Released into the wild, it is highly likely that SKYNET altered its properties and transformed the ground truths it was trained from. Unlike traditional kitchen strainers, algorithmic sieves have desires or emergent properties built into them, in which the so-called weights are altered either through evaluative loops, through tinkering and experimentation by humans, or through their own design via self-learning—so-called unsupervised machine learning techniques—or through so-called reinforced learning. That is, the original ground truths will be transformed by these algorithms through their epistemic operations, making it likely that the SKYNET program altered the thresholds of enemy and friend.

In addition, it is also likely that this would alter the very evidence that targeting analysts look for in their attempt to infer enmity, which would eventually produce targets beyond the similarities to the original "known couriers". Algorithmic sieves are then not only agents of interpretation but meaning-making actors in which the measure of valuation and thus meaning is the percentage of correct classification. However, they are also inherently contextual, temporal, and spatial, depending on the aggregated and disaggregated data and where in the wild they are inserted.

Much more is unknown about SKYNET, but it was not used as a closed-loop automated system directly linked to strikes like Igloo White was. Because it was interested in couriers and because couriers are more valuable alive than dead, it is likely that SKYNET was used to churn out selectors or seeds that would be further investigated through contact chaining and put under PoL persistent surveillance to sweep up the data exhaust these individuals may put out. It is thus unlikely that this was a system designed as a substitute for human cognition but, rather, was used as a perceptual, mnemonic, and epistemic aid or tool to alleviate human workload and provide insights from data that humans could not. However, through its workings, SKYNET would also contingently evolve, shaping and transforming the role of human cognition in the assemblage.

It is, however, clear from these slides and from other documents emanating from the NSA that the belief in these machine learning systems' ability to accurately pinpoint enemies and make networks visible based on metadata was very strong. Thus, these systems were part and parcel of and directed and drove targeting processes, not only in the sense that they provided leads and actionable intelligence but in that the different models created through these systems were combined and interacted and shaped each other, creating a meta-model in which the ability to interpret the outputs was often sacrificed for the presumed accuracy and enabling factors of these models.[98] Regardless of these problems, these algorithmic systems, were, in the eyes of the martial apparatus, highly valuable, and more and more individuals became followers of the "Church of SIGINT."[99]

The Gatherers

While the Church of SIGINT saw a steady growth in believers, the emphasis placed on the exploitation step in the F3EAD system is interesting, not only because it points in another direction to the SIGINT guided "collect it all" mantra but because it emphasizes that F3EAD, like the Phoenix program, is not just about killing. The idea was that killing the enemy would not lead to greater effectiveness against their networks, simply because capturing, humans as well as things, objects, artefacts, documents, and so on, yields more data, which again keeps the targeting cycle running. Capturing the enemy for purposes of interrogating (leading to TIR—tactical interrogation reports) is normally the preferred option, which also has the allegedly potential added value of capturing important documents and digital devices at-site for analysis (so-called document/media exploitation, or DOMEX), adding further layers of INTs to the already growing multi-INT family.

As the US targeting doctrine explains: "Documents and pocket litter, as well as information found on computers and cell phones, can provide clues that analysts need to evaluate enemy organizations, capabilities, and intentions. The threat's network becomes known a little more clearly by reading his email, financial records, media, and servers. Target and document exploitation help build the picture of the threat as a system of systems."[100]

To the martial apparatus, such smaller elements are not pieces of evidence in themselves but are an entry point, a "selector" or "seed," to find connections that can become a part in the heterogenous data assemblage that allows for navigation across and the intertwining of disparate elements. The doctrine goes on regarding interrogations of captured suspects: "The tactical questioning of detainees is crucial to revealing the threat's network. The ability to talk to insurgent leaders, facilitators, and financiers about how the organization functions offers significant insight on how to take that organization apart. Intelligence from detainees drives operations, yielding more detainees for additional exploitation and intelligence."[101]

US military intelligence officer Charles Faint and SOF officer Michael Harris argue that to achieve this, "the inclusion of law enforcement personnel and their investigative, forensic, and information-sharing capabilities were critical in the process of turning intelligence into evidence."[102] The bottom line of this "exploitation-analyze" step is to capture, gather, examine, and evaluate information and rapidly turn it into actionable intelligence. The emphasis on learning through gathering means that targets are not simply objects to be killed but also intelligence sources through which one can, the belief goes, gain additional insights.

A highly classified military report on ISR from 2013, obtained by the *Intercept* and released as the so-called drone papers, shows the problems the US martial machinery has conducting warfare according to F3EAD in areas where US military forces have limited human resources to draw on (e.g., Horn of Africa and Yemen). Indeed, as one of the slides clearly states, "assassinations are intelligence dead ends" that provide little follow-on data that can be exploited.[103] This is especially true of so-called drone strikes. Not only are the targets themselves not able to reveal more information about the network, either through tracking them or through interrogation, but the all-important multi-INT exploitation process is also severely limited, as there is no access to the documents, electronic devices, and so on that special operations raids would perhaps yield and that the ops/intel feeding process demands.

As one of the slides laments, there is "very little 'finished based' intel (DOMEX or interrogation) to drive next 'F3' cycle ... disrupting / slowing the 'cycle.'" The next slide goes on to discuss how traditional HUMINT and SIGINT can be "gap-fillers" absent TIR/DOMEX but argues that these sources are not as timely or immediately relevant for the continuation of F3EAD. In Iraq in 2007, TIR and DOMEX "from Finishing actions provided the bulk of phone numbers [above 80%], locations [approximately 100%] and terrorist names [80 %]" for furthering the cycle and to finish the next target.[104]

The importance of capture raids and so-called follow-on exploitation for the F3EAD is reinforced by a report on US SOF from one of the leading think tanks in the US, in which the authors state, "As SOF have improved their collection and interrogation capabilities, the bulk of new targeting information is generated by intelligence recovered during raids or through the interrogation of detainees."[105] While much has happened in the SIGINT world since then, this clearly shows the importance of SOF capture operations vis-à-vis airstrikes to keep the cycle running and allegedly learn about the enemy.

This does not, of course, mean that targeting is now less violent. On the contrary, John Nagl's famous statement that JSOC "is an almost industrial-scale counterterrorism killing machine" that conducted "roughly 3,000 operations in . . . 90 days" makes it clear that contemporary aspects of targeting are indeed very violent.[106] With roughly 30 operations a day, kicking in doors, blowing up walls, dragging people out of houses, and killing suspects at a frantic pace, the special operations forces left individuals for dead and families and communities in misery. And this is without counting the numerous of air strikes conducted despite limited actionable intelligence or the uncountable psychological effects of living under the potential for violence.

The importance of gathering intelligence and of capturing over killing is, however, an insight that stands in contrast to much of the critical literature on contemporary Western warfare, and in particular, drone strikes and targeted killings. Being mostly focused on the serious shortfalls and fallibility of the portrayed effective, efficient, surgical, clean, and virtuous warfare that the West legitimizes its operations through, the literature does not adequately take into account the aspects of targeting and warfare that fall outside of the killings.[107] The result is a tendency to overlook the emphasis on learning through operations as a very important shift in contemporary Western warfare, with its own serious consequences.

The immense tempo at which JSOC conducted its operations meant that the violence was different from mere killing. It used violence as a means to learn about the operational environment and the enemy network, something that brought the war rather indiscriminately into the everyday life of ordinary people on an unprecedented scale.

To Kill or Not to Kill

The symbiotic relationship between operations and intelligence created a particular martial dilemma between developing sources and raiding targets. While these dilemmas had long been known, particularly by the CIA in its small-scale violent manhunts around the world, for the military, this was something new. Some of the so-called difficult tradeoffs were certainly alleviated by the belief in the capture-exploitation-analysis equation, but it was far from certain that a raid would lead to the capture of either individuals for interrogation or any sources for DOMEX.

While F3EAD and the ops/intel fusion largely created this problem, it also, paradoxically, was believed to be the source of the solution. According to Lamb and Munsing's study of the SOF CT network, the fusion allegedly improved decision-making: "When operators know the value of a source, they can better judge the value of an operation that compromises the source, and they can be more appreciative of the need to collect certain types of intelligence. Similarly, if intelligence analysts understand the operations their analyses support, they can better tailor them for relevance."[108] At least in theory. According to the same study, "A constant tension existed between the SOF desire to hit targets as soon as possible and the Intelligence Community's predilection to protect sources and collect information for as long as possible."[109] The tension between killing or not killing was a constant problem that not only concerned the actual decision to kill or not to kill, but that clearly shows how the self-referential feedback loops engineered into the F3EAD processes drove thinking and practice.

As former CIA analyst and targeter Nada Bakos explains with regard to the differing cultures and approaches to the insurgency in Iraq between the CIA and the SOF, "It was starting to become clear that devotion to absolute speed at these raids compressed the OODA loop, thereby de-emphasizing analysis of the intelligence and undervaluing the effectiveness of the mission."[110] With the introduction of Flynn and McChrystal, and thus also the fusion of intel/ops and the experimentation with the F3EAD methodology, she argues, "the pace became much more frantic."[111] "SOF was taking a more horizontal approach, looking for insertion points where they could find them. Boom-boom-boom: they were daisy-chaining, grabbing a player and then going after the next viable target that guy knew."[112] The reference to the famous observe-orient-decide-act (OODA) loop is important here, as it clearly shows the emphasis on speed.

The OODA loop was created by US Air Force Colonel John Boyd in the 1950s to explain how decision-making occurred through recurrent processes of observe-orient-decide-act. His argument was that in a competitive environment such as warfare the goal should always be to go through this cycle faster than the opponent—to know its rhythms and OODA faster than them. By acting faster than the opponent, the argument goes, one could perpetually lock the enemy into an OO loop in which it was unable to make decisions and act.[113] Boyd's theory was, however, largely developed through experiences of air-to-air combat and not the highly complex environment of an assumed insurgency, which calls for deliberation and judgement rather than speed.

This was "the SOF military-style vision," Bakos argues. Blinded by speed, this "led to their misunderstanding individuals' roles within Zarqawi's network—if those individuals were part of the network at all."[114] The CIA, Bakos claims, had a different approach, wanting to keep "bad actors in place for the time being so they could inadvertently continue providing the CIA with valuable intelligence,"[115] always asking, "If the person was merely an interlocutor, might the target be of more use to us in place so that we could monitor whom he talks to and where he goes?"[116]

This claim about a specific military culture with a preference for speed and action is also one that John Lindsay picks up on. In his ethnographic account of the everyday life of targeting, while working in the service of Navy SOF in Iraq, Lindsay argues that the SOF culture was one in which the appetite for the offensive was so strong that it had turned the war into blind targeting practice, with the result being that tactical military operations often worked against the larger political strategy. According to Lindsay, the culture of the Navy SOF was reflected in the targeting process. The processes, from producing or ingesting data to analysis and interpretation of output, he shows, are all based on interpretive choices and prior decisions. Primed to construct and look for targets, the processes were biased toward action, toward the production of actionable targets, which, according to Lindsay, led to the detriment of a deeper understanding of the social environment.[117] In this way, the targeting process was a product of the military culture that reproduced itself via novel epistemic practices and techniques, creating self-referential feedback loops that were not ops/intel generated but came from their own biases and action-oriented habits.

While both Lindsay and Bakos argue that it was the culture of the SOF that led them to go on the offensive, in which speed was of the essence, the very assumptions and logics inscribed into the F3EAD methodology demanded that action be taken. F3EAD was designed based on the postulation that actions would generate further data to be exploited, thus making finishing operations a violent necessity. In this way, it is not only that F3EAD inscribed and intensified already existing cultural practices such as the appetite for the offensive and the desire for speed but that it was a technical device or methodology that had a tendency, and perhaps with purpose, to "unsettle discursive formulations and shake up cultural practices."[118]

There is also another related reason for the martial apparatus being so eager for action, related to F3EAD as a learning cycle, but that stems from the problem of so-called low-signature targets. In a world of presumed low-signature enemies, F3EAD extends its operations into the unknown to seek out these enemies. This is based on the idea that complex systems (e.g., presumed terrorist networks) often require some form of action in order to force target elements above intelligence thresholds. Reliant on an emergent understanding of the enemy and the operational environment, the military see the activity of probing as vital to making things emerge.

Probing-Learning-Perpetuating

Although saturated with different digital epistemic practices and infrastructures, the F3EAD methodology finds its logic not only in the systematization of the enemy and the search for making the unknown known but in operationalizing emergence through the systematization of processes of trial and error to learn about the operational environment and the enemy network. The hypothesis is that "by perturbing the target system, some target elements must necessarily respond, thus providing a signature that can be detected, analyzed, and turned into intelligence concerning the target system,"[119] which in turn is based on a belief that the threat is "out there" but has not yet fully formed or even emerged.

So rather than a counterinsurgency logic that "aims to pre-empt or prevent an insurgency forming,"[120] perturbation is akin to Massumi's idea of "incitatory power," where "you go on the offensive to make the enemy emerge from its state of potential and take actual shape,"[121] contributing to the emergence of the threat. Eyal Weizmann notes the similarities with Israeli military practice and incitatory power. In a footnote, Weizmann explains how this mode of action is thought of as tools for research. As the infamous Israeli general Shimon Naveh explained to Weizmann, "'Raids are a tool of research ... they provoke the enemy to reveal its organization ... most relevant intelligence is not gathered as the basis upon which attacks are conducted, but attacks become themselves modes of producing knowledge about the enemy's system.'"[122]

This, it can be argued, sets current targeting approaches apart from the notion of preemption and risk management logics.[123] Rather than trying to anticipate, deter, or manage the feared effect of something happening, this type of targeting cycle actualizes the potential of the threat to which it is ready to respond. In this sense, the idea is to engage in processes of emergence, modulating an action-reaction cycle in order to condition the enemy's reaction based on a preplanned script of future predictions.[124]

It is a mode of operation that seeks to transform the relationship between the present and the future by acting on something to make it materialize. Rather than just anticipating and preempting the future based on prediction, the emphasis is shifted toward creating a future through participation. While warfare has always been about creating a future, with F3EAD and the data logics through which it actualizes its operations, targeting is not simply planning, anticipating, responding to, or preempting, but actively seeking out opportunities through constant perturbations and feedback loops to learn.

Australian military theorists Anne-Marie Grisogono and Alex Ryan elaborate on this through what they call "adaptive campaigning."[125] Arguing that "hostile forces" will "largely be operating below the discrimination thresholds of surveillance systems," their proposed solution is to "take action first in order to stimulate reactions from which [friendly forces] can glean some information." Forces would have to "fight for, rather than with, information" and therefore have to engage and interact with the environment "to be prepared to continuously evolve their understanding and their approaches as they learn."

Turning John Boyd's OODA loop around, they go on to argue that "acting" is the first prerequisite to "stimulate a reaction" that can be picked up by human or technical sensors "from which to learn something."[126] "Adaptive action," as they call this feedback system, is defined as "an iterative process that combines the process of discovery . . . and learning"[127] with action. Critiquing effects-based planning and operations for having a too traditional and linear way of thinking about causation, Grisogono and Ryan argue that in complex environments it is impossible to start with an effect and discern from that the course of action. It is only through probing actions that one can stimulate responses through which one can learn about the context or test assumptions and then have an effect.[128]

The consequence of this design is that F3EAD not only uses violence to incite and produce an enemy but also perpetuates warfare in the sense that every potential action is also a potential source of data ready to be picked up by "the right sensing mechanisms already deployed, cued and ready to collect"[129] and churned through advanced computing systems. Not only are targets produced and selected for the perceived threat they pose, for their systemic functions, or because of their capacity for intelligence exploitation, but now operations and actions are themselves seen as means for actualizing unknowns. In this sense, F3EAD is aimed at triggering the affective and cognitive sensorium of the enemy to stimulate a reaction that can trigger one's own epistemic operations aimed at not only conditioning the enemy's emergence, per Massumi, but more importantly, to learn through constant operations.

This adds another layer to the perpetuation and endlessness of warfare that both de Goede and Weber have argued largely came from the network as devices or the data logics through which the disposition matrix is designed and engineered. The cyclical nature and the idea of perturbation build into F3EAD show that it is designed for sustaining operations indefinitely through learning.

In this sense, wars are not becoming perpetual or endless only because the so-called liberal world is unable to imagine conclusive endings, as Caroline Holmqvist argues[130] or because they are becoming increasingly global and permanent, drawing on policing logics to manage populations, as many critics have noted.[131] Rather, this is similar to the argument made by Nordin and Öberg that the inherent processual logics of targeting make war into a process ad infinitum, turning war into warfare and warfare into endless targeting practices.[132] The notion of incitatory power within F3EAD and the data logics through which it actualizes its operations and its feedback loops adds another detailed dimension to this argument.

While there certainly is an increasing entanglement of police and military thinking and practices in contemporary warfare,[133] the martial origins and design of targeting, with its emphasis on probing and perturbations, draw less on police, in the Foucauldian sense, as a biopolitical tool for the management of populations and creating order[134] than it resembles an instance of *cynegetic* power—the dyadic relationship between the prey and the hunter.[135] At the same time, however, the emphasis on learning means that the F3EAD methodology seems to be about more than the administration of death or hunting individuals or the return of the sovereign disciplinary power. While this can be seen as the "culmination of a longer modern history of warfare inflected with developing histories of scientific knowledge production and practices of killing and control,"[136] the contemporary targeting assemblage's emphasis on learning through action represent a break from past operational logics and rationalities.

The "Closed World" and the Expanding Battlespace

If one sees the F3EAD methodology as a machinery of learning, it resembles in many ways the conclusions drawn from HES and what Edwards and Knorr Cetina have located in their respective studies on computers and machineries of knowledge production—that they often become self-referential systems orienting themselves to the inside rather than the outside environment.[137] Whereas the Cold War enemy was bounded in time and space, the processual creation of the enemy as relationally networked nodes and links creates a different sort of closed world. F3EAD certainly relates to its outside environment in its encounters with the enemy and participation in emergent processes, but the cyclical nature of the targeting process is indeed a self-referential space where every bit of information feeds back into the knowledge construction machinery.

It is, however, less directed toward a central struggle than toward learning, carving out its own space rather than being guided by political or institutional orders. Military targeting at this historical juncture, is thus less of a product of some macro-political rationality to (re)produce order, as Edwards argues, nor is it solely a rational bureaucratic management instrument of policy as we saw in the case of Vietnam. Rather, military targeting has become first and foremost an end in itself, endlessly learning and translating the world into network models and nodes to perpetuate the cycle or "execute the F3EA[D] algorithm *ad infinitum.*"[138] The result is an ecology of operations trapped in feedback loops, thereby also perpetuating warfare in its wake.

F3EAD thus produces not only a closed world but an expanding one. It produces a different military spatiotemporality than territorial geopolitics, which is more bounded by sovereignty and physical geography. In this sense, it is the individual constructed target that defines the battlefield or battlespace rather than the other way around. Warfare has become target-centered as opposed to geo-centered, yet geolocation is key to the production and operationalization of the enemy through the data logics.[139] While producing the enemy is reliant on geotagging, warfare becomes dislocated from territory because the contemporary target assemblage and its TNA techniques and practices demand it.

This echoes some of Gregoire Chamayou's conclusions about contemporary warfare: "The boundaries of the battlefield are not determined by geopolitical lines, but rather by the location of the participants in an armed conflict."[140] Chamayou sees this shift as an instance of cynegetic warfare, arguing that the world has been turned into a hunting ground rather than a battlefield. With this hunting ground, he argues, there is no longer any reciprocity, no combat. And if there is no combat, there can be no war as traditionally understood. War, Chamayou claims, has now become hunting.[141] While these claims are exaggerated, especially in light of the above-described analysis of F3EAD and its focus on violent ground operations, learning, and probing, it is clear that contemporary warfare has fragmented, dispersed, and distributed the spatial dimensions of warfare at the same time as the temporal delineations have become blurred through different speeds, intensities, and perpetual targeting cycles. F3EAD thus adds further empirical insights to the many studies of drones and their abilities to channel warfare across space, arguing that the spatial and temporal delineations of warfare have become blurred, not only endless, but also everywhere reappearing in novel and terrifying ways as unevenly distributed "punctuated spaces."[142]

This does not, however, mean that the battlefield or battlespace has disappeared or that it simply has become a hunting ground. As an "imaginary arena,"[143] the battlefield reappears through targeting as a spatiotemporal network model[144] populated by objects, links, and relationships between individuals, places, events, physical objects, and infrastructures. This means that the battlefield is not only where kinetic violence materializes, or where it ends. Contemporary targeting with an emphasis on an emergent understanding of the enemy through operational learning brings warfare intimately into to the social fabrics and connective tissues of everyday life from which the network model of the enemy draws its data, creating an ecosystem of objects.

Conclusion

In the global war on terror, experimentation was the name of the game, networking was the buzzword, and teams of SOF, intelligence analysts, forensic experts, political analysts, cartographic and mapping experts, drone operators, computing and data handling experts, sensors, computers, databases, techniques, processes, methods, and models were assembled, configured, and wired into hi-tech networked fusion centers that would produce and operationalize the enemy. Nodal analysis would link suspects together to form a network of enmity, and PoL analysis would determine the suspiciousness and actionability of specific objects. "Advanced" interrogation techniques would be used to force captured individuals to give up leads, while captured documents, phones, laptops, and so on would be turned inside-out for clues. In the air, drones would sweep up communication and provide live footage of large areas, and computer systems would ingest and process massive amounts of metadata. On the ground, teams of SOF would design operations to stir up threat networks to make targets emerge from their sanctuaries.

The fusion of intelligence and operations in F3EAD and the configuration of an increasing variety of heterogenous elements would make sure to keep the targeting cycle running, turning it into an organized learning instrument, probing the environment and churning out targets at a frantic pace. With the presumed ability to uncover the unknown knowns and the known unknowns, the US martial apparatus reconstructed the battlespace and engendered a specific form of warfare. "Targeted killing," Hayden argued, was no longer kept in the confines of paramilitary organizations like the CIA but had "become a core part of the American way of war,"[145] a way of war constantly churning out the wrong people and, in its wake, piling up dead civilians and shattering societies.

Despite being armed with a novel targeting methodology, advanced sensor technologies to harvest data, and novel computational tools and techniques to process and analyze the data, the idea that the unknown unknowns were out there, waiting to emerge, troubled the US martial community. Not only did presumed "terrorists" look like everyone else, but they also behaved similarly to everyone else. That is, the activity patterns of a vast number of individuals showed so-called behavioral "signatures" associated with "terrorist" patterns of life, making it difficult for the US military to distinguish friend from foe and produce any type of knowledge about the enemy. The US martial apparatus determined that the problem was that their predefined signatures turned up either too many suspects or could not find those unknowns who did not exhibit their predefined signatures of enmity. They were certain that the "unknown" was out there.

The imagined solution to this irritant was born from the epistemological problem of the unknown unknown and emerged in the laboratories of the war on terror to complete the two-by-two chart from the introduction. As the next chapter will show, it was a solution that sought to shift the focus away from monitoring known locations or searching for known signatures toward discovering anomalies. It would come to mark a profound change in martial epistemology and what is of interest to targeting, and at the same time redefine and reconstruct the operational environment and transform warfare and its long-standing teleological conceptions.

6

Tinker, Tailor . . . Trial and Error: Experimental Design and the "Algorhythmic"

When you think about it, everything and everybody has to be somewhere.[1]
— James R. Clapper, former Director of National Intelligence and the NGA

Battlespace of Things and Gateway to the Real

I walked the analysts through the CSG's RT10 tool suite, first explaining how the new tools would let CSG[2] analysts call-chain their targets to identify their contacts, then enrich the data from all the databases in SHARK-FINN,[3] and finally display the data as a crisp, clean network in Analyst's Notebook.[4] What once took weeks now took a single analyst minutes to produce. Showing them IED-network insurgents moving across the big-screen GeoT map[5] in near real-time, Arabic text messages scrolling across the bottom of the screen as the insurgents sent messages from one target to another, I could see it was starting to sink in with many of the "ops" guys. "Yes," I said, "it's Enemy of the State, but now it's the insurgents who target the Coalition who now have nowhere to run."[6]

This vibrant description of a day in Baghdad from a National Security Agency (NSA) Cryptologic Support Group (CSG) deployer provides scintillating details of the supposed capabilities of the RT10. He is recalling a day he was sent to brief, and boast about, the capabilities of the RT10 to a mixed audience coming from different INTs and so-called other government agencies (OGAs), a military euphemism for various "civilian" security agencies such as the CIA and FBI.

The RT10 was a precursor to the real-time regional gateway (RTRG)—a data processing and data mining system introduced in 2007 that was configured to enable the real-time production and hunt for signatures and to enhance the capturing of feedback from probing operations. The RTRG architecture grew out of the NSA/National Geospatial-Intelligence Agency (NGA) GEOCELL program encountered in the introduction. In 2003, the NGA had been elevated to a three-letter acronym, from its previous four-letter acronym, NIMA (National Imagery and Mapping Agency), signaling the new-found importance of geospatial intelligence.

GEOCELL was the initial effort to fuse geospatial intelligence (GEOINT) and signals intelligence (SIGINT) together as complimentary forms of multi-INT and was driven by the perceived need to shift from describing large-scale activity to more fine-grained pattern-of-life (PoL) analysis of individual entities, in order to "track 'em and whack 'em." According to a joint press release, the goal of this configuration was "Horizontal integration—working together from start to finish, using NGA's 'eyes' and NSA's 'ears'—makes it possible to solve intelligence challenges that cannot be resolved through separate application of normal GEOINT or SIGINT methods."[7] GEOCELL was to combine what is often, in the martial community, termed "phenomenologies" into a hybrid system to enhance the targeting process.[8]

Although this resembles the idea behind Igloo White of combining different sensing apparatuses, the GEOCELL symbiosis was of a different breed. Unlike Igloo White, which was blind due to the absence of imagery, GEOCELL had an additional role in directing fire or ordering kill/capture operations based on tracking real-time movement. It was in charge of producing a number of analyses and activity reports in order to identify patterns and track people through space in real time and improve knowledge and understanding of the overall operational environment as well as providing support to soldiers on the ground. As a former NGA analyst recalled, "The work [GEOCELL] did in prosecuting the Iraq War was unimaginable. The effect they had on intelligence production ... made an immediate impact."[9] The apparent success of GEOCELL created further demand for computational systems that could store, mine, and process all the information created. One of these systems was the RTRG.

According to James Heath, the NSA director's science advisor, the RTRG was built to get "NSA cryptologic capabilities to the military front lines in a matter of seconds and minutes"[10] by integrating signals intelligence for geolocation of individuals. It was a system that constantly sought to produce and process data about everything and everyone all the time, which, in effect, brought the "collect it all" stature to the frontlines.

According to a top secret STRAP[11] page by the UK Government Communications Head Quarters (GCHQ), the RTRG is a data-centric program positioned as a so-called edge node on the overall SIGINT architecture and is tailored to the local environment. It relies upon

> an array of sensors streaming full-take collection into a forward-deployed regional data repository in parallel with offline forwarding to national

systems. . . . To fully exploit voice intercept RTRG is trialing new applications to perform keyword based selection and alerting. RT-RG provides access to tactical data sources in theatre via semi automated and manual data ingest into a local RT-RG repository. Federated query tools support multiple simultaneous cross-referencing of all these data sources, tactical and national, against incoming real-time intercept. These facilities offer huge productivity gains for analysts in rapidly *assembling a complete, fully informed picture of their target.*[12]

To support this fantasy, the RTRG was built as a comprehensive but distributed computing system that relied on numerous computers and software applications positioned locally to collect far greater volumes of data than was previously the case when information was sent back to the US. One of the primary benefits of this, according to James Heath, is that "collection that would otherwise have been discarded because it could not be directly linked to a known target is now subjected to analytic algorithms that can reveal new targets of interest."[13] In other words, data that were not immediately needed would be held in a database and be included in the massive amounts of heterogenous information that data analytics could sieve through, aggregate and disaggregate, and fold and unfold in an ongoing process in the belief that data would somehow solve the martial problem of the unknown. This obsession led to a staggering volume of data produced:

In order to provide a comprehensive real-time view of the SIGINT environment in Baghdad (a population of roughly 4 to 5 million residents), an NTG [national-tactical gateway, another precursor to RTRG] must be able to ingest 100 million call events [information or metadata about an intercepted phone call] per day, plus 1 million voice cuts [the content of the phone conversation itself] per day. In addition, it should be sized to ingest 100 million Digital Network Intelligence events per day. At this level of performance, an NTG can reduce traditional SIGINT timelines from 1000 minutes to 1 minute, an improvement of three orders of magnitude.[14]

The system ran on a hybrid architecture pairing the RTRG with what was called the GHOSTMACHINE, a cloud architecture that consisted of the open-source software Hadoop, allowing for large-scale data preprocessing and analytics, connected to a massive storage and search engine named CLOUDBASE, which was built inside the NSA. This architecture saw that the preprocessed MapReduce analytic results were fed back into the RTRG relational database and vice versa, so that they could match and draw on each other's data. To make these databases work for targeting, the system's architecture was made up of and managed by numerous software applications: Goldminer for metadata search, GeoT geospatial tools, Agent Logic for alerts, Sharkfinn for selector enrichment, SKS as a report and documentation manager, and Panopticon for target management.[15]

The result of this database architecture and its software applications within the RTRG was according to the martial apparatus the enabling of "a set of graph/network analytics and target development analytics"[16] that provided "a streamlined, integrated workflow"[17] saving considerable work hours compared to the previous manual processes. In addition, it also enabled analytics not previously available, such as the ability to search and spatially visualize the operational environment populated with various objects, targets, SMS messages, calls, and other forms of activity in real time.

Figure 6.1[18] visually portrays the standardized process of targeting within this novel ecology of operation, moving from SIGINT and HUMINT and other forms of data (tips), through various computational and algorithmic processes adding geolocational and temporal data (target development), which leads to a geolocation of a particular target through which a kill/capture operation can take place (outcomes).

While this system in many ways worked similarly to SKYNET, based on the belief that the enemy and its targets would appear through the filtering of data noise through various algorithmic sieves, this was closer to a SKYNET on steroids. And, according to the leaked presentation on RTRG in 2011, the system worked: it boasted that it played a key role in "90% of all SIGINT developed operations," "leading to 2270 capture/kill operations," "6534 enemies killed in action," and "1117 detainees."[19] What these body count numbers do not tell us anything about is, of course, the accuracy of these classifications or the truthfulness of reports themselves, for that matter. Because it was designed to keep up the frantic pace according to special operations forces' (SOF's) vision of war and F3EAD's cyclical logics, and was not engineered for accuracy of its outputs, it is likely that thousands of civilians are included in these numbers.

Despite the many possible fallibilities of this system, which certainly are numerous, the RTRG configuration presented the military with a completely new way of producing and operationalizing the enemy *and* the battlespace, together at the same time, entangling time with space and matter, into a novel ecology of operation.

The result of these messy epistemic operations—ingesting, fusing, searching, and processing all-INT data—enabled and heralded, for the martial apparatus, a transformation in the production of the enemy, providing

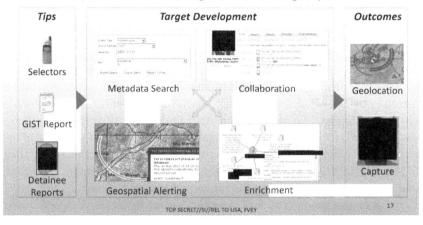

Figure 6.1

RTRG operational process. *Source*: National Security Agency (NSA), "NSA Presentation on RTRG Analytics for Forward Users," 16.

what the martial community saw as real-time actionable intelligence to the frontlines, that is, producing and making available a real-time updated model of the enemy and the operational environment. In addition to this supposed real-time ability to search for and track targets, the RTRG also had the ability to ingest so-called blue force tracker data, showing friendly forces, so that the system became a virtual battle ecosystem or ecology of operations, which, according to its proponents, "greatly enhance[d] overall battle space visibility."[20]

The problem, of course, with such abstractions produced by multiple filterings and disaggregations of data is that these models, much like the HES, largely obscure the uncertainty on which they rest, generating not a "gateway to the real" but a model that obscures the "reality of the real," making some things visible while leaving other things out. Although the systems were designed for speeding up the targeting process, and not accuracy of the outputs, the potentials of the RTRG were not opaque to its users, stimulating the long-term martial dreams and fantasies about a transparent battlefield and automated knowledge production and decision-making.

In an interview in 2009 in *SIDtoday*,[21] Parker Schenecker, deputy NSA/CSS representative to US Central Command (CENTCOM), argues that they are finding broader applications for RTRG all the time, and the usefulness of the system "is limited only by our ability to wrap our head around its potential."[22] The RTRG would eventually be succeeded by DARPA's (Defense Advanced Research Projects Agency) Nexus 7, a cloud-based system that, according to the head of development, Randy Garret, included "essentially every kind of data there is" and expanded beyond data strictly relevant to the military to include patterns of everyday life, such as transportation and prices of food, and experimented with crowdsourced data to measure progress and make guesses about the future. As one of the programmers stated, speaking directly to Schenecker's comments about the potential of RTRG, "We were really using the latest research in quasi-experimental design, in machine learning, and data mining literally on hundreds of intelligence feeds to make inferences about what would happen next."[23]

In sum, the RTRG served as the all-encompassing datamining and data-processing system to accompany the NSAs "collect it all" mantra and the probing logics of F3EAD. While the RTRG created a technical illusion of progress with regard to the problem of the unknown, the belief that there were still things out there that could not be uncovered by searching for known signatures or that simply refused to be forced above the detection thresholds of the martial sensorium and probing techniques still haunted the martial apparatus. The solution to the unknown unknowns was to turn the traditional analysis upside down and reconfigure the martial epistemology of military targeting.

The Mysteries of the Unknown Unknowns – from Signatures to Anomalies

Rectified through years of experimenting with novel data analytics, computational logics, machine learning algorithms, experimental methods, and organizational reconfigurations and fusions, activity-based intelligence (ABI), we are told, emerged as a novel mindset and methodology in the early 2010s. ABI soon became the method of choice to uncover so-called unknown unknowns[24]—a method previously described by Derek Gregory as a sort of "militarized rhythmanalysis" or a weaponized time-geography,[25] after the French philosopher Henry Lefebvre and the Swedish geographer Torsten Hägertsrand-and would go on to produce yet another fantastical model of the world and the enemy.

ABI, like signature hunting, works from the assumption that objects of interest are only visible through their activity and that these are only discoverable through the careful tracking of movements and transactions. However, unlike signature hunting, ABI does not assume that there are such things as predefined signatures, a known, but, rather, that enmity is defined by deviations from the normal. The unknowns in this rendition are targets of interest that have no signatures to track or search for but that, according to its proponents, can be discovered through how their activities and transactions in space and time deviate from the normal.[26] Here, the normal would be redefined as the normal workings of everyday life, digitally remastered, and that which stood out from this normal, or anomalies in the computational martial language, was that of interest.[27]

This so-called discovery process is thus distinguished from other targeting processes by the absence of prior knowledge of the research object; there is no known from which to start the process. Rather than producing pattern-of-life signatures and ideal types of targets—be it missile launchers, underground bunkers, malicious codes, or individuals—and then searching/tracking for these signatures or profiles, like needles in a haystack, ABI targeting first builds the haystack through real-time tracking of the entire environment and then finds that which sticks out from this haystack. The methodology was thus based on the premise that both the haystacks *and* the needles are unknowns, but rested on the belief that if one was able to produce the haystacks, the needles could be discovered.

Although signature-based and anomaly-based targeting operate alongside each other, the rise of ABI signaled a shift in martial epistemology and what is of interest to targeting. Moving the focus away from monitoring known locations or searching for known signatures and toward discovering anomalies, ABI adds a distinct spatiotemporal dimension, not only to the operationalization of enemies and warfare but to the *production* of enemies and creation of worlds. How and with what effects are what this chapter now turns to.

ABI and the Anomaly

ABI is defined by the Office of the Undersecretary of Defense in the US as "a discipline of intelligence where the analysis and subsequent collection is focused on the activity and transactions associated with an entity, a population, or an area of interest."[28] According to former NGA director Letitia Long, ABI is used to "identify patterns, trends, networks and relationships hidden within large data collections from multiple sources: full-motion video, multi-spectral imagery, infrared, radar, foundation data, as well as SIGINT, HUMINT and MASINT information."[29] In these vast databases of geotagged data, the goal is to discover seemingly irrelevant patterns of life and anomalies in the observations, trends, and patterns by correlating all the multiple sources of information available, which without ABI's particular temporal, spatial, and network analysis would remain invisible. Or so we are told. Systems engineer William Raetz talks up the believed potential of ABI, stating that "ABI holds the same promise for imaging the Human Domain that the invention of the camera held for imaging physical terrain."[30] But ABI was not merely a camera picturing the so-called human domain; it was a very specific machinery to produce and operationalize the world.

In ABI this production takes place through novel data analytics that ingest, fuse, process and visualize multisource spatial and temporal data to make up an empirical "normal" of an entity, an area, a digital network, a system, or a population.[31] Although ABI was born from the epistemological conundrum of the unknown unknowns, its techniques grew out of the potentials of a number of other sociotechnical systems, like the RTRG, and the realization that when integrating data and information from multiple sources and source types, known as multilevel data fusion, the one thing these data bits had in common was locational data. So, with all data geotagged and databased, geography became the common dominator favored for exploring and analyzing countless pieces of information from various sources. Querying geospatial multi-INT databases with new information, analysts came up with the process of "geo-chaining," a technique to connect people, things, and events to specific locations. This, according to NGA's head of ABI, Dave Gautier, "evolved into the building blocks of ABI."[32]

The ABI methodology, we are told, rests on four pillars: "georeference to discover, integrate before exploitation, data (sensor) neutrality, and sequence neutrality."[33] Geotagging of all data, we are told, is important so that the data can be visualized both spatially and over time so that trends and patterns in the growing network start to jump out. The idea is that if you ingest data in bulk, you can build a database of datapoints, much like the RTRG, that might be relevant for the future. First, the more data are geotagged, the larger will be the increase in discovery potential and level of confidence in the results. Second, and importantly, documents disclose, all data should be fused and folded together before they are exploited, so as not to lose important small bits of data that might be relevant in the future. Third, all data are equally important, so do not be fooled by the flashy SIGINT data; "pocket litter" might be just as valuable, ABI proponents argue. And, finally, there is sequence neutrality, which means that the analysis should not happen in a predefined sequence from A to B but, rather, should let the data define the activity being sought.[34]

According to the champions of ABI, the reason for these pillars is to change the culture and mindset of traditional collection-driven intelligence, in which collection of data is undertaken to answer specific questions derived from hypotheses, as with, for instance, the Consolidated Target Intelligence File (CTIF) and the Bombing Encyclopedia (BE) database. In ABI, data are supposed to drive the analysis rather than the analysis driving the collection, making it, according to an industry presentation of ABI, deductive and nonlinear as opposed to inductive and linear.[35] It is thus an empiricist methodology in which flattening and folding all data to make it similar and computable in the same manner is of vital importance. The idea is that this not only allows for answering preexisting questions but also, allegedly, enables the discovery of entirely new questions based on interrelations deeply embedded within the data. In this respect, ABI is more akin to abduction, creating a hypothesis and making inferences about things it does not know.

From a scientific point of view, "anomalies represent a break in the pattern of normalcy pertaining to spatiotemporal patterns ... but they only exist secondary to establishment of normal patterns of behavior. Therefore, it is important to first understand the normal state of affairs of an environment and provide a proper logical and statistical classification schema to capture it and use it for computations."[36] Such logical and statistical classification schemas, using methods such as clustering, outlier analysis, classification algorithms, and statistical approaches, build the models of normality that are subsequently used to detect new patterns that deviate from these models.

Although there are, as Aradau and Blanke point out,[37] many different techniques for anomaly detection, producing different types of anomalies—point, conditional/contextual, and group/collective—these practices are usually referred to "as the process of detecting data instances that significantly deviate from the majority of data instances," drawing together various computational communities including data mining, machine learning, computer vision, and statistics.[38] The main problem with anomaly detection, from a scientific position, is that anomalies are, by definition, irregular. They are therefore rare data instances associated with many unknowns in behaviors, data structures, and distributions and are consequently difficult to discover.

While detecting anomalies in single data sources, also often referred to as outliers, has been a practice for decades, what is of particular interest to ABI is the area of spatiotemporal anomaly detection usually defined as the absence of spatiotemporal continuity.[39] These are time series analyses in which patterns and anomalies are based not merely on temporal dis/continuity but also on space. In other words, ABI is reliant on the data being not only time-stamped but also, as the first pillar of ABI demands, geotagged. ABI can thus be said to produce the category of conditional or contextual anomalies—individual data instances that are anomalous in a specific spatiotemporal context. To discover these anomalies, ABI is reliant on data from multiple heterogenous sources such as multidimensional data, graphs, images, text, and audio.[40]

The believed importance of spatiotemporal activity data is underlined by an NGA presentation on ABI showing how intelligence moves from pixels, or geographical locations, through movement, to rhythms, and then to network graphs and activity maps in which statistical distribution (the normal) can be used to discover and evaluate future values for anomalies (figure 6.2[41]).

According to Willian Raetz, ABI was partly designed to amend the problems of traditional social network analysis (SNA) or threat network analysis (TNA). The problem with network analysis was that the attributes of entities were difficult to analyze because the metadata about an entity are pinned to an individual object, making them difficult to cross-correlate. Instead, Raetz argues, ABI allows for creating "metadata objects and defining links between them," making it possible to analyze "common touch points," which adds new richness to the data. As such, Raetz argues, "by creating a time-derivative view of both the social and the attribute graphs the analyst can also receive information suggesting behavioural patterns over time."[42] Through the eyes of military targeting, ABI is thus a way to address the limits of both traditional network analysis and signature-based detection.

ABI is thus not another algorithmic practice that groups data according to the presence or absence of particular features, signatures, or profiles in

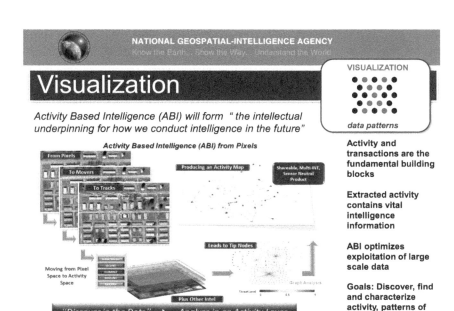

Figure 6.2

Activity-based intelligence. *Source*: Todd G. Myers, "GEOINT Big Data: Implementing the Right Big Data Architecture," NGA 2013 Joint GMU-AFCEA Symposium (2013), 10.

Tinker, Tailor . . . Trial and Error: Experimental Design and the "Algorhythmic"

the data. It is interested in difference, rather than sameness. We may think of the difference between anomalies and signatures as being "born" strictly within or outside of data. While anomalies are based on a strict empirical, which does not mean objective or unbiased, rendering of the normal and thus born strictly inside data, signatures are born outside data, in the military imaginaries or cultural norms of society. While signatures may be partially defined and set by data, as objects of scientific and statistical calculations, and thereby sociotechnically derived, such profiles are nonetheless culturally specific predefined categories into which humans are indexed. Signatures are thus closer to what Ian Hacking refers to as setting "boundary conditions" that are fundamental to governing populations through statistical laws.[43]

The key thing is that anomaly detection does not operate from any predefined norm, apart from the belief that the irregular is of interest. That which sticks out, is out of place, suspicious, or otherwise curious does not derive from a sense of norm but, rather, from deviations purely constructed from data. Such a belief rests on an epistemic logic that anomalies do not stand out of social order but stand out of data. A target is then produced not on the basis of the data of a person or object but, rather, on the basis of what algorithmic systems have learned about others in the environment. The "other as anomaly" is thus a result of "making up the normal."

While not superseding other ways of producing the enemy, the shift from predefined normative signatures to anomalies has profound consequences. In one account, it signifies a new epochal change, one that surpasses Foucauldian disciplinary society and the Deleuzian societies of control with its own distinct effects.

Welcome to Targeted Societies

In the article "Patterns of Life: A Very Short History of Schematic Bodies," Grégoire Chamayou contends that targeting is now so prevalent and influential that it "serve[s] as the effective operational basis for an entire series of practices of power." Chamayou's hypothetical that we are entering "targeted societies" rests on his analysis of the chrono-spatial techniques of ABI, which he argues, produces an "individual-dividual synthesis." This is a "form of individuality [that] belongs neither to discipline nor to control, but to something else: to *targeting* in its most contemporary procedures, whose formal features are shared today among fields as diverse as policing, military reconnaissance and marketing. It might well be, for that matter, that we are entering *targeted societies*."[44]

Chamayou argues that targeted societies work differently from Foucault's processes of "normation" and "normalization," which Foucault argued were part of the mechanisms of discipline and apparatus/*dispositifs* of security in his studies of the emergence of "governmentality." In the 1977–1978 "Security, Territory, Population" lectures, Foucault drew a sharp distinction between what he termed the disciplinary techniques of "normation" and the apparatus of security. In the first regime, a predefined "norm" is "fundamental and primary in disciplinary normalization,"[45] defining the normal and the abnormal in which the goal is to force reality in line with this norm. In contrast to the fixed norm of discipline, Foucault tells us, the apparatus of security is "exactly the opposite," in which "the operation of normalization consists in establishing an interplay between these different distributions of normality" based on statistical regularities and irregularities that serve to establish a normal and that are then deduced to the norm.[46]

ABI and its chrono-spatial techniques, Chamayou argues, challenge Foucault's notion of the apparatus of security, which rests on an understanding of normality based on statistical distribution of an average normal population. According to him, ABI does not produce normative normals, nor does it produce an understanding of normality based on statistical averages. In fact, as Amoore and Piotukh argue, contemporary data analytics and "*big data* marks a significant break with statistical notions of what is *of interest* or concern."[47]

Aradau and Blanke, offering a "thick description" of anomaly detection, argue that it has emerged as a technique precisely to address "the limitations of statistical knowledge and risk governmentality"[48] for security professionals and, as we have seen, the limits of signature-based detection and network analysis for the military apparatus. For them, "anomaly detection is indicative of the transformation of the algorithmic subjects of security, as it is equivalent neither to abnormality nor to enmity." Rather, they argue, anomaly "emerges as a supplementary third term" of interest, reconfiguring security dichotomies away from the traditional friend/enemy toward logics of similarity/dissimilarity. An anomaly therefore appears as an object of analysis on its own, not as an error but as a desirable result of analysis in machine learning.[49] It thus signals a move beyond the human recognition and interpretation of patterns of behavior as suspicions, something being out of synch, strange stuff, and so on and toward what is standing out of long-term spatiotemporal data capture and processing, rendered as different or dissimilar.

For Chamayou, the technical spatiotemporal production of individuals creates something much like the F3EAD approach: "individualities considered both as objects of knowledge and of intervention." But he continues: "The corresponding object of power here is neither the individual taken as an element in a mass, nor the dividual appearing with a code in a databank, but something else: a patterned individuality that is woven out of statistical dividualities and cut out onto a thread of reticular activities, against which it progressively silhouettes in time as a distinctive perceptible unit in the eyes of the machine."[50]

This "lashing of an aggregation of data to individually indexed chronospatial paths," Chamayou argues, thus also differs from Deleuze's prognostic of societies of control.[51] While one can say that the apparatus of security, as outlined by Foucault, responds to contingencies, fluidities, and complexities of reality, the goal of norm-setting and norm-production differs considerably from the logics and operation of targeted societies. Statistical aggregates or dividuals like those described by Foucault and Deleuze, while not necessarily working from a defined signature, operate based on a decision on what to measure. This decision on what to measure is thus made prior to an engagement with statistical numbers. The very operations of ABI, on the other hand, stem from the mysteries of the unknown unknowns and are powered by the "collect it all" and "analyze it all" mantras, in which there are no prior decisions on what counts as important or not. While this, of course, is an assumption and based on a choice, it is not based on a norm of what counts as interesting but, rather, is driven by the mysteries of and belief in the unknown.

Targeted society's individuals are thus not simply new modes of descriptions demarking the boundary between normal and abnormal made possible by rendering people as objects of scientific inquiries and statistical calculations[52], but something else. As Chamayou argues, these chrono-spatial anomalies, or what he, after Paul Klee, terms "dividual-individual synthesis," are "in a sense, *normativities without norm*. Their notion of the normal is, in fact, strictly empirical: it is *learned* by the machine on the basis of an analysis of frequencies and repetitions in given sets of activities. It is then a discrepancy with such patterns of regularities—an *anomaly*, rather than an *abnormality*—that will trigger the red-orange alert on the analyst's screen.[53]

In ABI the focus on rhythms over singularities, the relations, and transactions between units rather than the units and their links, is central. Repetitions in ABI are understood as spatiotemporal patterns that produce ordered space-time, but repetition also produces mutations and difference. As with the rhythmanalysis of Henri Lefebvre, the relationship between repetition and difference is crucial. It is, namely, through repetition that difference is produced; anomalies, or objects of interest, thus emerge from the act of repeating, a sort of excess produced out of the normal repetitions of daily life.[54]

This is why, Matteo Pasquinelli states, the "two epistemic poles of pattern and anomaly are the two sides of the same coin" in data mining. According to him, this creates a situation in which the "construction of norms and the normalisation of abnormalities is a just-in-time and continuous process of calibration" in which the abnormal resurfaces as "an abstract and mathematical vector"—the anomaly of computational rationality.[55] Anomalies can only be detected against regular patterns, and the result is that anomalies do not refer to a "normatively inscribed deviation from the normal [but rather] what is simply irregular existence."[56]

But what does this shift from normative signatures to anomalies without norm mean for the practices of power, security, and warfare? For Chamayou and his notion of targeted societies, these techniques are not constructed in order to bring the deviant in line with the normal: "These devices have no particular model trajectory they seek to impose on the various lives they monitor. Their normativity without norm is animated by another goal, another kind of devouring appetite: to spot discrepancies in order to 'acquire targets,' and this in a mode of thought where, targets being unknown, it is the unknown that becomes targeted."[57]

Targeted society is not interested in normalization or the construction of norms, and it does not tell you how to behave according to a norm, because what is "normal" and what is "anomalous" is relational; they produce each other through real-time data processing. Neither is it interested in control through goal-directed feedback as in cybernetic logics. Targeted society is also much narrower than biopolitics and societies of control that encourage self-governance via contingency management as a principal means of securing the well-being of populations. It operates through a different logic—a logic that is aimed at acting on and with emergence through an experimental logic of trial and error, intensifying and authorizing specific modes of attention toward difference. Ironically, Chamayou contends, it is thus the very notion of liberal life, the sacred individual freedom to follow one's own path, to be different, that immediately is being signaled as suspicious. The result of ABI was that the martial apparatus now considered both similarities, as in signatures, and differences, as in anomalies, as suspicious, leaving very little room for anyone to navigate the violent sensorium of military targeting.

From Targeted Societies to Military Targeting

While Chamayou links ABI and anomaly detection techniques to a whole new knowledge/power regime, in which targeting in all its forms operates as a basis for an entire series of practices of power, from a *military* targeting perspective, the story is different.

It is very much the case that targeting, in both language and operation, is omnipresent and creeps far beyond its martial origins, entangling seemingly disparate arenas from cybersecurity, global health, urban spaces, and police practices to capitalist markets. The list of things with targeting attached to them is long—targeted sanctions, targeted adverts, targeted development, inflation targeting, political targeting, behavioral targeting, targeted therapy, targeted immunization strategies, targeted drug delivery, microtargeting, critical infrastructure targeting, targeted surveillance, and targeted killings. While we can say that targeting, and, in particular, the spatiotemporal analytics of ABI, serves as an effective operational basis, or the conditions of possibility for a variety of practices of power, as per Chamayou, treating all forms of targeting as the same borders on technological determinism; emphasizing technological operations over content and context fails to recognize the specifics of its various logics and the situatedness of its applications.

Although many forms of targeting operate with and through many of the same data-processing technologies and techniques, seeking out patterns and anomalies, objects and relations, military targeting is distinct because it is designed from and works through specific military imaginaries. And it is the martial epistemologies that provide a framework for the design of the configurations that shape what is of interest to the input-output processes. Different purposes and different uses based on different contexts, cultures, imaginaries, assumptions, and the environment, into which these configurations are inserted, produce and construct different objects of interest and capacities to know and act.[58] In short, and to channel Hayles again, the context in which the interpretation and judgement of data is done matters.

Churning the same data through a military targeting assemblage and a marketing assemblage would not produce the same outputs, nor would they necessarily be interested in the same inputs. There is still a difference between targeted adverts and targeted killings. Military targeting is not interested in singling out individuals according to riskiness or threats from a statistical aggregated mass but, rather, classifies and selects them according to their relational value to the overall model of the enemy. It is thus not first and foremost interested in single entities but in the spatiotemporal relationships between areas, people, and objects, providing data to the modeling techniques. It is through these relationships that it produces its operational model of the enemy and finds its nodes and links, which, when severed, would have the most disturbing effects on the whole. And it is these whole-part relations that distinguish military targeting from, for instance, marketing and its production of "algorithmic identities,"[59] which are based on signatures of similar digital behaviors, or the dividual-individual synthesis that Chamayou identifies. As Aradau and Blanke point out, "the logics of war, risk and data produce specific modes of otherness." The military enemy is different from the "risky" migrant and the threat signatures of cybersecurity companies.[60]

If we return to the notion of epistemic operations, we can see how context matters in the situated process of interpreting information. It is context that connects information with meaning. As computational machine learning techniques have recast anomalies as the desirable result of analysis,[61] targeting configurations are now engaging in a much more experimental production of the enemy in which both patterns and deviations are things that can be further probed. The key shift is that these epistemic operations do not start from an assumption of normal/norm, or what counts; the notion of what matters is inscribed at a later stage in the manufacturing process. This alters the situated epistemic process of interpreting information.

While previous targeting configurations would inscribe imaginaries, assumptions, norms, and/or meaning into the very design and engineering of the systems—the way sensors were set up to and trained to produce specific data, and the particular processes through which data were analyzed and archived—anomaly detection systems mainly inscribe meaning at a later stage. This does not mean that the way the systems are designed, data is labeled, or the specific techniques employed for detecting anomalies are not part of assumptions, but that these do not conform to a predefined notion of what is of interest, apart from everything being of interest.

Unlike signatures, anomalies do not contain meaning other than being something irregular in a larger datascape. If a signature pops up on a screen, the analysts know what it is because they have been trained to recognize what the computational systems have picked up and because the systems have been designed to look for specific things. If an anomaly stands out, there is no immediate answer to what it is because it is, by definition, an irregular occurrence, an unknown. So, while the detection of anomalies can be automated, the interpretation and meaning-making related to them happen at a different temporal moment.

Because these systems are interested in the mutations and difference that naturally occur in the processes of rhythmic repetition, the output is always an unknown and a suspicion, not a ready-made piece of actionable intelligence. It is at the output stage that a form of "normation" comes back in to infuse meaning into the anomalies. These are particular operations conducted by trained targeting analysts employing military imaginaries, along with human gut feeling, and certain biases related to expertise, experience, and training in order to either confirm the anomalies as suspicious or simply deem them irrelevant and incorporate them into and update the normal. Such human interpretation is complicated by the fact that the logics and algorithmic models used to compute the normal and the irregular favor what, in the computer sciences, is termed *detection accuracy* over interpretability and explanation of the identified anomalies.[62]

The focus on detection accuracy is part of a shift in production of the enemy that assumes objects not to be homogenous and stable, but to be constantly mutating and shifting according to the repetition and difference that processing of spatiotemporal data from the rhythms of everyday life engenders. Through the techniques of object-based production (OBP), objects are now literal placeholders or buckets for all information and intelligence produced that can be relationally linked to that object, be it an individual, a thing, or a place. And by extension, these objects are also launching points to discover more information. As such, they work as a hinge around which data, abstractions, and outputs dynamically form and unfold through recurrent epistemic operations.[63]

Interpretability is seen as less important, as the anomaly is not only a way to get at that which is unknown and suspicious but is increasingly viewed as an opportunity to be further probed and exploited. ABI is not interested in producing truths but, rather, in producing clues that can be operationalized in order to modulate affects, learn, and engage with the emergent operational environment. The result of ABI completing the two-by-two matrix is thus a highly fluid ecology in which data are constantly ingested, processed, folded, and stored and (re)processed in large cloud-based databases. Here, the figure of the enemy and its targets is in constant becoming. It is a phantasmic model of the enemy that emerges only through these configurations and epistemic operations and where targets are always partial and constantly updated as the computational systems learn through de- and reaggregation of data.

A Database of Past, Present, and Future

Rather than producing a fixed ontology of the enemy with an essence, in this computational regime, the enemy is produced through chrono-spatial computational logics that render them as anomalies. It is a sort of target individuation through "technogenesis,"[64] where the enemy is always emergent, in formation, always existing in potential in the relation between the constantly updated chrono-spatial model, ongoing data feeds, and machine learning algorithms.

Since anyone has the potential to become a target sometime in the future, Chamayou contends, archiving the microhistories of individual lives becomes paramount. This, according to Chamayou, is an apparatus that forms the "instrument of *biographical power*," creating sleeping dossiers on individuals in case they become interesting.[65] While such sleeping dossiers, or target packages, do not necessarily differ from the vast numbers of targets found in the Bombing Encyclopedia or the disposition matrix, how they are produced and constantly (re)made and for what purpose differ. But, as we have seen, these dossiers are far from sleeping; they are wide awake, living in massive databases, constantly being (re)figured and revisited as new information enters the systems, making sure that everyone and everything has already been targeted and are constantly retargeted. In doing so, targeting produces a fluid and ever-changing enemy model that is only visible through the computational rendering of its past and present activities and transactions.

Here, we may think of Luciana Parisi and Steve Goodman's concept of *mnemonic control*, in which "virtual memory, the entanglement of the past and the future in the present, [becomes] the locus of preemptive power."[66] According to Samuel Kinsley, drawing on Berndt Steigler's notion of mnemotechnics, the "industrialization of memory" through large-scale software programs, databases, and algorithms, has "significant agency in the various ways in which we collectively communicate and remember . . . constituting novel sociotechnical collectives which have begun to influence how we perform our lives such that they can be recorded and retained."[67] This is not just the mnemotechnical ordering of discrete information that freezes the world at a particular moment as, for instance, per the BE and its rendering of the enemy as infrastructural systems ripe for destruction or the more static profiles or signatures that form the basis of F3EAD.

Importantly, it is "*also* concerned with the routinisation of activity, of constituting rhythms by which activities are organized and comprehended"[68]—a continuous record of spatiotemporal processes. This facilitates an intervention into our ongoing spatiotemporal experience that is "reshaping the process of retention," thereby producing "access to 'what happens.'"[69] According to Wolfgang Ernst, what we are witnessing is a move from the old traditional passive archive and the purely technical practice of data storage without a narrative memory to more active databases. It is a shift from archival space to archival time in which the dynamics of permanent transmission and processing of data produces as much as it archives events, where data models on the inside dictate the narrative.[70]

What we can glean from these systems of retention is that events, or what happens, become predetermined by the very encoded rules of the targeting configurations and their epistemic operations. That is, the configurations condition and control events in the sense that their retentional activities produce the present and the future in specific ways—the temporal folding of the past into the present to anticipate the future. In this way, the "past stays in potential, continuously ready to actualize its present,"[71] not in linear continuity or separated from the future but, rather, in a way where past and future coexist in the present as a model of normality. This is a rationality that operates through an unlimited self-propelling dynamic of collecting and analyzing the "archives of a future past."[72]

Here, ABI operates as a multitemporal situational awareness where "presentism" is constantly informed by the past and where databases of the past are given new relevance by the present. Through these epistemic operations, fragments of the past are constantly reordered in light of new data, in which neither history, the present, nor the future is transcendent, shared, or standardized but, rather, are moments of relation and remnants of different windows of time that conjoin and produce particular correlations of the normal. And it is through these moments of relation that the enemy and targets emerge—as momentary arrhythmias to the rhythms, things that deviate from the empirical normal and become signs or indications of enmity.

Algorhythmic Targeting

It is useful here to think about ABI techniques as "algo*rhythmics.*" An algo*rhythm* is, according to media theorist Shintaro Miyazaki, "the result of an inter-play, orchestration and synthesis of abstract algorithmic and calculable organisational concepts, with rhythmic real-world signals, which have measurable physical properties." It is a neologism made up by *algorithms* and *rhythms* that "points to the very processes of computation itself . . . and manifest[s] itself as an epistemic model of a machine that makes time itself logically controllable and, while operating, produces measurable time effects and rhythms."[73] The concept is useful here, as it sensitizes the analysis toward a specific engagement with the time-based character of computing at the same time that it allows for injecting temporalities into the epistemic operations of the configurations. As such, it can enable a better understanding of how time is entangled with space and matter in relation to their constant processing of rhythmic signals and information from their environments and their objects, processes, materials, and bodies.

Seeing ABI as algorhythmic, we can say that it both functions as an analytical technique to process rhythms of planetary life to produce the enemy as spatiotemporal patterns and its deviations *and* as a way to produce real-time management and interpretation of feedback loops. Also, we can begin thinking about how this algorhythmic rendering of the enemy not only brings into being a novel kind of enemy and targets, but through its entangling of time, space, and matter also produces distinct forms of military space-times and warfare.

Claudio Coletta and Rob Kitchin adapted the work of Miyazaki and developed the notion of "algorhythmic governance" to investigate the ways in which codes and big data are deployed to regulate temporal rhythms of urban life.[74] Where Coletta and Kitchin depart from the increasingly vast literature on how algorithms are used to actively monitor, manage, and control populations[75] is in their conjoining of both the spatial and temporal work of algorithmic governance. Reframing the temporal dimension of algorithms and their work as algorhythms, they argue that algorhythmic governance is "structured forms of knowledge designed to create eurythmia."[76] Seeing algorhythmic governance as part of a broader shift in urban governmentality enacted by and through algorithms, Coletta and Kitchin argue that governmentality is shifting its focus from disciplinary forms to those of social control, from molding and restricting subjects within spatial enclosures like the panopticon to seeking "to modulate affects and channel action across space."[77]

What they are interested in is how technologies are being used to mediate and regulate rhythms, that is, seeking to "limit arrythmia and produce eurythmic systems."[78] Smart cities, Kitchin explains, are "space-time machines, with networked infrastructure and smart city technologies significantly disrupting temporality as well as spatiality to produce a new set of space-time relations."[79] Here, time becomes a resource in which the "availability of real-time data seemingly creates an annihilation of space and time,"[80] much like the truncating of the so-called military kill-chain and targeting's transformation of war into endless operations on a global battlefield.

Although not considering the novel role of algorhythmics in producing different kinds of Others, like Chamayou and Aradau and Blanke, Coletta and Kitchin's notion of algorhythmic governance nevertheless directs attention to the importance of thinking of time and space together and the enabling of channeling "action across space." This speaks directly to the way in which the battlefield has reemerged as an abstract model of military space-time,[81] or what we can here call the "algorhythmic battlespace"—a complex space-time ecosystem made from a limitless set of heterogenous rhythms and patterns.

Returning to the work on real-time smart cities, one can think about this emerging battlespace in terms of "smart" and the relentless work of turning the world into a giant sensor, or what one security analyst has called "an unimaginably large cephapoloidal nervous system armed with the world's most sophisticated weaponry."[82] This is akin to what Benjamin Bratton has referred to as "the stack," an "accidental megastructure" that is both a computational architecture and a new governing architecture.[83] What algorhythmics adds to these spatial metaphors is that it inserts time into space and adds rhythms to architecture, so that we can speak of an algorhythmic military space-time. Conjoining the time-based model of algorhythmics with the spatial metaphor of battlespace merges aspects of timing and process together with structure and architecture to capture how the operational environment is being reshaped by martial epistemic configurations that ingest, process, and act on real-time rhythmic data.

Whereas battlespace is the organization of space remade through martial means, an algorhythmic battlespace is a mathematical space-time typology constructed from and through abstract algorhythmic operations and material reality. Organizing and arranging matter and time in space, algorhythms also generate patterns, rhythms, affects, and effects, thus modulating specific courses of action. Rhythm-making in this battlespace is, however, different from the traditional military battle-rhythm that seeks to dictate tempo and control rhythms according to a concrete planning methodology to minimize arrhythmia (disruptions) and produce eurhythmia (order). Operations in the algorhythmic battlespace seek to become with the rhythm of the battlespace, and, in contrast to algorhythmic governance, the martial apparatus is less interested in seeking out or regulating eurythmic orders than it is interested in creating, experimenting with, and learning from arrhythmia in order to harness its potential.

With the US martial assemblage now seemingly able to monitor, research, search, and discover on a planetary scale in accordance with the analytical two-by-two matrix, the NGA outlined a grandiose and fantastical vision to create a continuously replicated virtual version of the entire world, a "digital twin," to operationalize all of the knowns and unknowns. This massive data ecosystem would not only transform ecologies of operation but also allegedly enable the martial apparatus to seek out emergence and immerse itself with the algorhythmic battlespace.

The Map of the World

In its *Future State Vision 2018* strategy, from 2013, the NGA laid out a plan to integrate all the diverse data and INTs to provide the foundation for the layering and synthesizing of intelligence to create a "common, integrated intelligence picture...accessible, anytime, anywhere." This was to be a customer-friendly database and dynamic map, referred to by the NGA as the Map of the World (MoW), which was to provide no less than "a unified, online, geospatial, temporal and relational view of the world."[84] Fueled by the craze of big data tools and techniques, and with an abundance of data ingested from an increasingly ubiquitous sensorium, the MoW would now serve as the foundation for multi-INT integration, or as the "geospatial bedrock" for all knowledge, information, and intelligence.[85] The MoW was thus to function both as a future cosmological imaginary of all the knowns and unknowns and as an imperative for further investments, practices, and materializations.

Edward Abrahams and his GEOINT colleagues argue that the "persistence and depth of data now readily available allows new products to be woven together out of three basic threads or classes of information: vector-based knowledge (everything), locational knowledge (everywhere), and temporal knowledge (all the time)." "This extends the analytic paradigm," they go on to argue, "to a knowledge environment in which every property of every entity is always available in real time ... represented as a vector of its attributes—i.e., all the metadata about that entity." As almost "every entity on Earth beacons vectorized metadata into a ubiquitous data cloud," the authors foresee an immediate future in which the whole earth is blanketed and indexed like a "spinning Google Earth globe with an infinite number of layers and an infinite level of detail"[86] across vector, location, and temporal domains. In this digital platform twin of the world, the fantasy continues, analysts can immerse themselves in data about everything and everyone in every location, all the time and across timespans, providing an update on the RTRG collect-it-all ecosystem for the targeting community to experiment with and exploit.

This Map of the World, we are told, is thus not only a 3D blueprint of the world's terrain and buildings. It contains much more than merely the static objects that used to make up most of the Bombing Encyclopedia of the World. The MoW is, in the words of three GEOINT specialists, a "living, global, multidimensional digital twin, [a] constantly replicated virtual version of the world" with "an activity-based intelligence layer tracking individuals and populations over time and cross-referencing their actions." Everything, from the "accumulation of 3D data on our bridges and houses, the tracking of our movements via GPS or wi-fi or health-monitoring devices, the thermal assessment of our bodies in public spaces, the nature of our very genes—are all being aggregated in one database or another as individual puzzle pieces and trends,"[87] forming the contingent evolution of this digital twin of the world. By creating "a unified ecosystem of multimodal data [as a] multidimensional atlas of humanity,"[88] the authors continue, it also serves as an ultimate historical archive and mechanism to record and assess our species and the landscapes we inhabit. Clearly alarmed by what they see as an inevitable entanglement of different digital representations of time and space into the digital twin of the world, the authors warn that "we find ourselves not just on the cusp, but already wading into a technocratic variant of a temporally and spatially tracked society."[89] This is a caution that finds affinities with the notion of targeted societies.

While Chamayou's general claim about targeting ushering in an epochal societal change is a provocative and novel analysis, it does not recognize that the objects of interest of *military* targeting go far beyond individuals. As NGA's Map of the World is a testament to, military targeting has become more all-encompassing than the form of individuality produced by ABI. Rather than solely a reductionist position of abstracting from the world—"replacing the part of the universe under consideration by a model of similar but simpler structure"[90]—the Map of the World becomes an endlessly updated gigantic complex ecosystem configuring a symbiosis between intelligence and operations, knowledge and actions, enemies and operators, and humans, and machines. Unlike the abstractive practices of the natural sciences intended to discover "objective truths" about the world, the MoW is not interested facts, but rather, is obsessed about how to actionalize the world.

In this cyber-physical conjunction, a plurality of different and disparate data layers and worlds entwines in a complex spectacle. The multitude of realities and perspectives in the Map of the World, its proponents claim, accommodate for error, ambiguity, and paradox by materializing them into something useful, something malleable and manipulable, lacing the world together into a more graspable and actionable model of the whole. What emerges out of this claim to apprehend the entirety of the world is not chaos or disorder but, rather, an experimental world of causalities and correlations. By creating a synthesis of a multiplicity of visions and proximity zones between disparate phenomena, it is claimed, the MoW, with its inbuilt functionalities and logics, frame the relationship between space, time, and memory through which the enemy can be recombined and remodeled and made actionable. If this sounds abstract and messy, it is because these technical abstractions are messy, partial, incomplete, and faulty and far removed from the horrors and reality of war. These are practices that rest on the long-standing martial dream and fantasy of automating perception, memory, knowledge production and decision-making, but with consequences for those on the ground that are all too violently real. However, this is a mess with a specific purpose for the martial apparatus.

This is a military space-time in which targeting works algorhythmically to produce the enemy in spatiotemporal experimental ways enabling a becoming-with warfare. In the martial sphere this becoming-with is thought of as "immersion." "By immersion," former director of NGA Letitia Long states, "I mean living, interacting, and experimenting with the data in a multimedia, multisensory experience with GEOINT at its core."[91] Through a global network of sensors and platforms, the NGA envisions a future that is "persistent, immersive and anticipatory," placing the analyst "at the center of the data," "living there every day" in the digital twin earth of the Map of the World. This vision comes close to Norbert Weiner's notion of the whole Earth subjected to future technopolitical and computer regulated control: "To see the whole world and to give commands to the whole world is almost the same thing as to be everywhere."[92]

However, Dave Gautier, NGA's activity-based intelligence lead, argues, "I think immersion goes both ways. We need to have our network and our algorithms immersed in the knowledge that the analysts have in their minds about our intelligence problems."[93] What this is alluding to is the concept of "computational thinking"—a notion put forward by computer scientist Jeanette Wing to describe the skill set of being able to formulate a problem and express its solutions in a way that can be computed and carried out by humans, by machines, or by cognitive assemblages.[94] For the NGA this means the "ability to think like a coder." It is about understanding the way machines work and how data are queried in order to maximize the potentials of computers and to shift the role of humans towards answering "why" questions.[95] The goal is the automation of abstractions, where *abstraction* is the formulation of a problem, whereas *automation* is the expression of a solution.

By seeking to immerse humans with the MoW and algorhythms with the minds of the analysts, the NGA is seeking to configure a particular cognitive assemblage. It is an assemblage in which the MoW cannot simply be thought of as a "container of possibility" for perception, memory, and knowledge but, rather, must be understood through immersion. When immersed, the hope is that the operators and analysts do not simply perceive the world but, rather, processually experience it. Being immersed in data-worlds provides a particular mnemonic for the full spectrum of human-machinic experience in which the immediate, physical, emotional, and unconscious response to the world, rather than careful deliberations, is what drives the actualizations of potentialities. With the belief that many different preexisting potentialities exist within the MoW, the martial hope is that the world can be actualized and experimented with in various different ways.

One way to think about these shifts is through the works of British mathematician and process philosopher Alfred North Whitehead.[96] If we, as Whitehead, think of the world as interrelated processes, the immersion within data-worlds and its effects can be thought of as a process of becoming "concrete," or actual. *Concrescence* is the name Whitehead gave to the process of becoming concrete, or novel togetherness, with which he emphasized how things, objects, and data are grasped to become actual.[97] In a Whiteheadian sense, it is the processes through which the "subjective" activity of immersion experiences the MoW that makes potentialities "objective." Seen through this Whiteheadian process theory, the experiences of immersion in data worlds make targeting into a particular aesthetic form of reasoning, emphasizing the immediate sensory and unconscious response to the environment rather than careful and drawn-out deliberations. In other words, through immersion the world does not become concrete as a representational picture of the world but is, rather, grasped and presents itself as immediate, momentarily flashes of actuality through the analyst's process of becoming-with.

This immersion within data-worlds is, of course, radically different from Donna Haraway's usage of "becoming-with,"[98] but it does point toward the martial hope of transforming the analyst's relation to the environment, one in which knowledge is felt rather than thought, not isolated but, rather, involved in an infinite network of data and signals. In this view, the world is no longer thought of as composed of discrete things but as something akin to Karen Barad's agential realist ontology, as "phenomena-in-their-becoming"[99], where both targets and operators become objects of the ecosystem. This is a martial epistemological framework in which the spatiotemporal position of things in complex ecologies is not learned through abstract knowledge, replacing a part of the world with a model of a simpler structure, but, rather, is experienced in situ through the affective capacities afforded by immersion within the MoW.

Although the MoW and immersion certainly open up the question of agency within the configuration, they also add another dimension to Lucy Suchman's notion of "contemporary bioconvergence,"[100] where both the targets to be captured/killed, destroyed, or otherwise manipulated and the operators are fused and locked together in specific ecologies of operations and cognitive war-fighting assemblages. The lofty hope for the martial apparatus is that this will alleviate the epistemic ruptures that occur with different situated knowledges.

Although these novel reality formations operate alongside older ones, they arguably represent a shift in rational and operational logic resting on what Luciana Parisi calls a "transformation in logical thinking activated with and through machines"[101] alongside simultaneous shifts in politico-military discourse and practice. While novel assemblages of heterogenous epistemic practices and material infrastructures have engendered new ways of making up people, society, and the planet, military professionals and policymakers have, at the same time, become more attuned to and comfortable with a world that is allegedly increasingly complex, unpredictable, and full of contingency—a world, it is claimed, that cannot be secured, governed, or engaged with in traditional ways but that has increasingly given way to the idea of governing emergent effects in real time.[102]

This form of sociotechnical governance, it has been argued, is not interested in solutions but represents a new form of management seeking to adapt and respond, rather than control and direct the world.[103] While this calls into question epistemological and ontological assumptions about how we know and how we govern and secure the world, it largely reflects the shifts in targeting that this book has already outlined: toward learning through doing, in which unknown effects become the driving force of (re)configurations, (re)learning, and (re)action. This "doing," it will be argued, manifests itself in the increasing entanglement and convergence of the experimental ethos between computational tinkering and tailoring and the operational trial and error ethos of so-called military design thinking.[104]

Techno-Rationality and Computational Experiments

Critical scholars seem to be in concert that big data analytics, machine learning techniques, and so-called artificial intelligence are central to the way in which the world is now perceived, known, and thus acted upon. Contemporary data science, Louise Amoore tells us, is interested in correlations over causations and has moved away from probabilities to possibilities, with profound consequences for the governed.[105] These world-making operations have increasingly come under scrutiny, focusing on "the manifold ways that algorithms are imagined, enabled, deployed, and utilized in practices of governance" and how they shape the exercise of power as a means to automate the disciplining and controlling of society.[106] Much of this literature calls attention to how the analytics are operating and their effects on social and political life and the nature of the governing and the governed.[107]

While datafying "everything" and putting it into databases for queries is, as we have seen throughout this book, hardly new, what is novel and important, according to Jutta Weber, is the shift from narrative to postrelational databases and flexible algorithms. This, she argues, defines the rationality in the present epoch of "technoscience," leading to a profound reconfiguration of epistemology and ontology in which technology assumes a principal role in addressing the most pressing of society's challenges.[108] But, as Amoore and Raley write, "it is not merely that algorithms are applied as technological solutions to security problems, but that they filter, expand, flatten, reduce, dissipate and amplify what can be rendered of a world to be secured."[109] In other words, the data ingested, the databases constructed, the algorithms used, and the world modeled are all important parts of the "narrowing of vision" that translates the complexity of the world into a "selective reality" that can be manipulated, controlled, and governed.[110]

Leo Breiman, the renowned statistician working on building bridges between traditional statistics and computer science and considered to be the father of the Random Forest machine learning technique, argues that there are two different cultures in the use of statistical modeling to reach conclusions from data. According to Breiman, the culture of traditional statistics, using methods such as "linear regression, logistical regression and cox models," "assumes that data are generated by a given stochastic data model" and fits the models to the data to draw quantitative conclusions. The problem with this, as we saw in relation to the HES, is that the conclusions are then "about the model's mechanisms, and not about nature's mechanisms"; the focus is on making the models fit the world and not the problem at hand,[111] leaving out such things as anomalies or seeking to incorporate them into the models. The benefit of such data modeling is that it allegedly produces a simple and understandable picture of the relationship between the input variables and the responses but, according to Breiman, limits understanding or outright produces faulty understandings.

The "algorithmic modeling culture," developed outside the field of statistics in computer science, treats the data mechanism as an unknown. Through various machine learning techniques and algorithms such as neural nets and decision trees for fitting data, we are told, algorithmic modeling shifts the focus from data models to properties of algorithms with the aim of predictive accuracy. As Breiman explains, "The approach is that nature produces data in a black box whose insides are complex, mysterious, and, at least, partly unknowable. What is observed is a set of x's that go in and a subsequent set of y's that come out. The problem is to find an algorithm f (X) such that for future x in a test set, $f(x)$ will be a good predictor of y."[112]

Moving to an algorithmic modeling culture, Breiman argues, has the benefit of accuracy over simplicity, but as accuracy demands more complexity, these algorithmic models lose interpretability; "we are facing two black boxes, where ours seems only slightly less inscrutable than nature's."[113]

Echoing Breiman's so-called accuracy/interpretability dilemma, computer scientists Pang et al. state, "Most anomaly detection studies focus on detection accuracy only, *ignoring the capability of providing explanation of the identified anomalies*. To derive anomaly explanation from specific detection methods is still a largely unsolved problem, especially for complex models."[114] In these systems, it is not only the complexity of the world, which is allowed for, as with the HES, but also that the complexity of the algorithmic models is welcomed. That is, machine learning models are appreciated for their ability to provide accurate and simplified outputs with limited preconceived notions of what matters, regardless of the inability to interpret and understand their complex operations. Thus, it is not so much that this narrowing of vision is novel, but what is novel is the way and the tools by which the complexity of the world is constantly under translation and the acceptance of so-called accuracy over explanation.

To Rob Kitchin, this big data analytics amounts to a new scientific paradigm, a radically new epistemological approach for making sense of the world, serving as the basis of a new form of empiricism: *exploratory science*.[115] Jutta Weber argues that this is a form of techno-rationality that "draws upon a logic of tinkering that data mines the unknown, exploring all manner of (often highly unlikely) possibilities . . . to *produce* patterns of correlations between data and thus to 'discover knowledge in databases.'"[116] Unlike modern scientific rationality, she contends, echoing much of the empirical material analyzed in this chapter, contemporary technoscience is not interested in the intrinsic properties of entities but, rather, in their behavior and relations, shifting the emphasis from representation and understanding to investigation and intervention. Importantly, Weber goes on to argue, "the key features of technoscientific rationality include formalized and systematized tinkering as well as the use of trial and error, bottom-up search heuristics and post-processing for the solution of complex problems," making "adaptation, imitation and imagination . . . key to a rationality aimed at resourcing the unpredictable and to attempts to find ways of exploiting surplus processes by technical means."[117]

However, from Rosenblueth and Wiener through Hacking, Daston, Galison, Heyck, and others, this book has shown that scientific rationality was always interested in the interrelationship between representation and intervention and the manipulability of the abstracted object, in processes of emergence and in the unknown, whose epistemological logic is rooted in systems theory and cybernetics. Regardless of where one wants to land in the debates about scientific rationality, the important part is that a contemporary computational regime of machine learning techniques and neural networks represents a form of epistemology that is not problem-solving but, rather, is a general method of experimenting with problems that needs to be understood in terms of abductive reasoning. A form of reasoning that is interested in the creation of new explanatory hypotheses, in which "unknowns [acquire] an ontological superiority able to transcend the epistemological certitude of scientific knowledge."[118]

While both of Breiman's cultures can be said to be a generalized activity of commensuration—translating qualities into quantities according to a common metric[119] privileging black-boxed predictive power over interpretable model parameters, the algorithmic techniques at hand alter the way in which human-machine configurations generate choice and elevate the role of algorithms from directional and goal-oriented preprogrammed problem-solving entities to experimental actors and operators. Not only are machines involved in the shaping and construction of reality, but they are increasingly also epistemic agents determining possibilities, generating choices, and demanding actions. This is a point that speaks directly to Hayles's notion of cognitive assemblages and raises the question of how machine and human agency are formative of and are formed through targeting and how meaning is processed through the interpretation of information.

For Hayles, the developments of machine learning techniques and neural networks mean that machines and computational networks are no longer just processing numbers but must be reconsidered as true political actors, as they can create meaning in themselves, based on the learning abilities they derive from designing and engineering processes. Neural nets, of which so-called large language models (LLMs) are the latest craze, are always in a process of becoming through the data they work on, and by making these models of the world, they also generate outputs.[120] The condensation of multiple possibilities into a single output is important, as it is this output that becomes the point of decision-making and action. In short, as Louise Amoore contends, "the single output of the machine learning algorithm is rendered as a decision placed beyond doubt; a risk score or target that is to be actioned."[121]

What we see here is that this computational transformation has significant consequences—for how we can understand what the world looks like for these cognitive assemblages, for how these worldviews are constituted in relation to humans, but also, and importantly, for what it makes happen in the world. While ABI presents itself as a nonlinear, deductive logic, the machine learning "insides" of this targeting methodology have gone beyond the deductive. With the alleged limits of deductive reasoning in algorithmic modeling, this has become a departure point for more experimental tinkering in the determinations of actionable intelligence, which has arguably transposed beyond computers and led to a cultural transformation of reasoning with and through machines. If anomaly detection techniques and the algorithmic modeling culture in general favor so-called accuracy over interpretability, the question of how epistemic operations are connecting the interpretation of information with meaning becomes significant.

Although Amoore argues that the output from machine learning techniques is placed "beyond doubt," an anomaly in itself, as has been argued in these pages, is not a finished piece of actionable intelligence. It still needs to be put into context and inscribed with meaning, a labor that is largely entrusted to humans. However, this is an analyst that cannot know how this single output was rendered beyond doubt, as interpretability has been traded for accuracy. The result is a troubling but nonetheless intriguing paradox; through the operationalization of the mysteries of the unknown unknowns, the martial apparatus creates its own known unknown—a known deviation from the normal but about which knowledge of its insides is unknown. However, this is nevertheless the martial assemblage's own unknown, and an unknown that to them generates opportunity—opportunity in the form of potentials that can be probed and experimented with through military action.

In this "regime of computation" the logics of discovery and experimentation find affinities with a movement within Western military circles, namely, that of *military design*, a movement that embraces experimental visions of the world and the logics and operations of creative disruptions, rejecting end states in favor of constant engagements and future beginnings.[122] Although this community is still in its embryonic stage, its vision of the world as constantly becoming and emerging, in need of possibilistic logics and iterative tinkering, experimentation through trial and error, and learning by doing reflects the conjunction of analogue and digital forms of reasoning galvanized with and through machines. This has profound consequences for how war is imagined, assembled, and operationalized.

Warfare as Design – Creative Disruption

Military design thinking has its origins in Israeli military operational thinking[123] from the mid-1990s and has since then gained increased influence, spread, and mutated into many Western militaries over the past decades. According to its proponents, military design thinking "is an umbrella term for a more or less consistent assemblage of reflexive approaches including complexity theory, systems thinking and postmodern social theory."[124]

Co-opting a variety of theories, military design thinking critiques the traditional natural sciences approach to war that seeks to churn out general knowledge and principles in order to produce a linear planning logic from model to particular application to effect. It starts from the assumption that the problem or failure of contemporary military interventions lies in the epistemology of the military professions, the sort of technical, positivist, functionalist, objectivist, or rationalist approaches that poststructuralist and critical theory have long criticized, and which the earlier pages of this book have also swarmed over. Thus, this differs from traditional military critics of interventions, which most often take issue with the application of violence rather than its epistemic roots. Military design thinking, then, operates as a critique of operational planning and its built-in logics of predictability and engineering reason, much like this book has already argued.

However, military designers are not interested in critiquing the military, militarism, or war as such but seeks ways to remedy what they see as the failings of orthodox military theory and planning techniques. It is interested in finding solutions in a world that is allegedly more complex than before. The challenge of contemporary complex conflicts, they argue, "demands a radically new thinking at odds with rationalism" that emphasizes the need for reflexivity about knowledge produced and used, focusing on problem-setting rather than problem-solving.[125] Rather than reproducing modernity's promise of predictive planning and control, military design promises to provide the martial apparatus with a novel way of thinking and particular techniques to govern complexities and manage uncertainties and the unpredictability of war's generative and messy nature. It thus offers a particular orientation that shapes how the world is sensed, perceived, and understood, with its very own violent operations.

At its core, military design thinking is about creating a mechanism of self-disruption for the military to go beyond doctrine, which according to Ofra Graicer, one of the most prominent military design thinkers, will lead to the ability to reinvent "itself through experimenting with emergences."[126] Continuous self-disruption is thought to be a way to both undermine traditional understandings and to create new understandings. The goal is the transformation of military institutions into perpetual learning systems, which, we are told, increases strategic understanding and operational efficacy.

While its theoretical interventions famously draw upon so-called French poststructuralism and architectural theories on deconstructivism,[127] military design also draws on chaos and complexity science, which was first made popular in the 1980s by US Air Force Colonel John Boyd[128] and in the 1990s through the so-called revolution in military affairs (RMA) debates. Military designers are, however, not interested in finding a "hidden order"[129] that could be used to overcome the fog of war and predict outcomes but, rather, use these theories to argue that if everything is connected and complex, what is required is an approach to warfare that is experimental rather than planned. As one military professional has argued, "military units should create 'portfolios of strategic experiments' . . . to go about the environment conducting experiments. This could simply be gathering information, it could be *acting 'on the system'* and then observing the system's response, or it could be something else. . . . The main function of higher headquarters would be to develop processes that would encourage a variety of action and figuring out how *to gather feedback in order to choose the most promising 'experiments.'*"[130]

Operationally, this is similar to what was argued in chapter five through the notion of probing, perturbation, and incitatory power, disrupting the system through experiments in order to capture feedback. Military design is thus not about governing complexity, because what vitalizes practices such as ABI is chaos. Chaos is what causes rhythms, rhythms generate data, and data is the fuel through which learning can take place. It is a form of action akin to what artist and social analyst Hito Steyerl calls "creative disruption, fueled by automation and cybernetic control."[131] According to Ofra Graicer, creative disruption "plays on exploiting tensions in the observed system that alter it and create a new one. . . . In that sense, strategy exploits or leverages tensions; tactics oppresses tensions; and operations mediate it into a system. We should think of the operation as a domain that mediates tensions, leaving the command system a *flexible space to deliberate, interpret, experiment, and make mistakes.*[132]

Graicer's statement resonates with Claudia Aradau's findings from her studies of intelligence and secrecy in "the age of big data," where she argues that it creates a system in which "it is impossible to think the error of knowledge. Error does not undermine the production of knowledge, but is integrated in the production of knowledge."[133] Error thus becomes part of the experimental tinkering and is internalized as a way of learning. The emphasis on learning through error is a different mode of epistemic intelligibility that is based not on past knowledge but on data, past and present. But unlike in the sciences, an error in warfare means that innocents are killed.

Understanding knowledge and strategy as being constantly in flow, Graicer argues, "In ecological terms strategy is always constantly changing and context dependent. In other words, strategy is first and foremost the potential to transform, in four dimensions—your way of thought, your understanding of the world, your organization, and only then—the world itself."[134] Design as creative disruption speaks directly to and upholds the idea of war's generative powers[135] and transformational potential,[136] although in a much less linear cause-and-effect and deterministic way than traditional military theory. In the world of military design, there is no longer instrumental certainty to planning and warfare, only violent experimental tinkering.

Although there is an ongoing debate within the design community between the purists who would like to thoroughly disrupt traditional planning and those seeking to incorporate design thinking into existing planning methodologies and practices[137], Dan Öberg sees design discourse as indicating a "possible shift from an imagery relying on tactical missions to counter chaos, towards warfare becoming part of an experimental vision of the world."[138] Such a shift certainly increases the well-known problem[139] of strategies turning into tactics and war into warfare, through endless tactical operations.

What military design does, however, is not critique the lack of strategic focus over tactics but argue that endless tactical-level operations are experimental and creative and thus important for understanding "systemic emergence"; operations are, as we have seen, vital for learning. However, this requires a different skillset aimed at constantly reconfiguring elements and the world to create new functional relations, new meanings, and new values. Military design is thus about the violent creation of functional environments through which targeting and war can flourish, and with which the martial apparatus can immerse themselves, and become-with, to form, deform, and transform that very same environment they have created. Violent operations are now fundamental to martial knowledge production.

This experimental vision of the world arguably stems from a rejection of end states and an embrace of "future beginnings ... which attitudinally indicates a need for constant engagement, reflection, and updated appreciations of how a system continues to evolve in response to the actions being taken."[140] This rejection of possible end states is less a strategic reflection based on assumptions that the West finds itself in a protracted intractable conflict and more of a reflection of the way newfound epistemic practices, material infrastructures, and data logics have transformed military thinking toward ideas of a world in constant becoming and emergence, which are supposedly in need of possibilistic logics and iterative tinkering and experimentation through violent trial and error.

Although emerging before the contemporary computational era, military design thinking has in recent years found increased convergence with computational logics and the ideas of immersion that reside in the MoW. Not only does the notion of military design rest on the ability to see things in their emergence and thereby establish a need for data analytics and other computational tools to capture operational feedback loops, but military design and computational sciences find each other in the use of experimenting and trial and error to redefine problems and generate knowledge about the world. This is a shift that does not simply concern technical apparatuses but, as Parisi argued, is a symptom of a transformation in logical thinking activated with and through machines. These are logics that have moved beyond the notion of control and involve a mode of reason that has accepted radical uncertainty and contingency for which the solution is a constant involvement that is geared toward exploiting anomalies and tensions as they arise from data.

In this world according to military targeting, the enemy is no longer a structure, a system, or a network but an unknown future event in the algorhythmically produced Map of the World ecosystem, where everything and everyone, everywhere and all the time, are not only potential targets for action but have already been targeted and are constantly retargeted.

Becoming-with the Algorhythmic Battlespace

Embracing radical uncertainty and contingency and an "end to ends" is a rationale that is affirmative to the Rumsfeldian unknown unknowns in which threats are always out there, in potential, but have not yet fully formed or even emerged. As the previous chapter showed, the idea of the permanent presence of a threat in the background logically calls for a new way of operationalizing the threat, a way of making indeterminate threats determinate by "perturbing the target system" in order to get "target elements" to respond in order to make them emerge above detection thresholds, which in turn can be made into intelligence concerning the target system.[141]

Threats, Massumi contends, only exist in the abstract, and thus, preemption is about setting in motion processes or tapping into emergence to which martial machinery is ready to respond. The idea is to engage in processes of emergence, or exploit tensions in the design language, to achieve *ontopower*, "a power through which being becomes."[142] There are, however, two crucial aspects of the contemporary targeting assemblage that differ from Massumi's theory of preemption and ontopower—the emphasis placed on harnessing the possibilities of emergence rather than closing them off according to a script and the emphasis on the normal/anomaly rather than threats/objects.

While contemporary targeting can be said to rest on the ontological premise of preemption, understood as when the nature of the threat cannot be specified,[143] and the epistemological condition of "objective uncertainty,"[144] this is no longer seen as a problem. Nor is setting in motion a prediction to foreclose it the goal of operations. The unknown has become part of the martial operational logic and is increasingly seen as a potential for knowledge; the only certainty is uncertainty, and the only way to know is to act, because actions generate data and data generate learning.

It is through the epistemologies of military design and advanced computational analytics that the current targeting assemblage finds its operational logic and practices. This points toward a novel condition of possibility to act upon or with the future(s). While this can be thought of in terms of incitatory power, operating as "agents of chaos"[145] or arrythmia makers, targeting not only forces unknown threats to emerge, but also seeks to experiment with different forms of future(s). That is, military targeting is interested not only in the emergence of threats, linked to specific actors but also in casting a wider net seeking out "future beginnings" and facilitating their occurrence through "creative disruption."

This does not mean foreclosing an uncertain, indeterminate future by producing it in the present but, rather, means seeking to *become-with* emergence, always experimenting with different possibilities. This is not about prediction, which relies on knowing how a system behaves, but rather, is about what military designers call "influenceability" or "nudging,"[146] which relies on knowing and learning how systems change and adapt—the study of transformation.[147]

One important aspect here is that the unknown does not have to "become" in the sense of being materialized or fixed, but is operationalized as always "becoming," empirically produced through making up the normal. This becoming, we are told, can be manipulated through the logic of disruption adding energy or information into the system while taking advantage of computational analysis and operational design.[148] "The role of the strategist thus becomes that of the acupuncturist—carefully analysing systemic flows, designing multiple and dynamic interventions across key intersections, ultimately building on the system's own complexity and momentum for its very disruption."[149] This, it is believed, offers a way of harnessing and utilizing complexity, uncertainty, and contingency for advantage.

In operational terms, this means that the military is not seeking to control or to limit effects through prediction but is trying to harness and tweak potentials that might or might not occur through operational probing—a sort of martial nudging that seeks disruptions rather than destructions of systems. It shifts the operational logics away from "predict to preempt" toward "anticipate to exploit."

As creators of chaos, the military now sees potential in war's generative powers as a way to learn about the operational environment and the transformational processes of emergence it both produces and allegedly immerses itself in. This focus on experimentation and repetitive design can be seen as a reaction to the presumed failures of preemptive anticipatory practices and is why the design movement advocates a move away from the military as a solutions-based organization and toward a learning-based organization that is always generating knowledge in and through experimenting—a becoming-with warfare.

In rhythmanalytical terms, the Map of the World creates an ecology in which human and computational cognizers[150] are immersed in data, seeking to measure and reveal the polyrhythms of the planet, and linking military operations as rhythm/arrythmia-makers to make the world move so as to be captured. Here, algorhythms are not only part of mediating and shaping the rhythms of targeting but are part of a so-called real-time military situational awareness. In this world, past, present, and the future are all present at once, evoked and utilized simultaneously, precluding more complex temporalities. But they are also shaping and structuring forms of data and thus also the forms of knowledge that enable real-time action and the radical (re)construction of military space-time as an algorhythmic battlespace—highly specific environments that the martial apparatus creates around themselves, and which they paradoxically depend on for their operation but also seek to deform and transform at the same time.

Much like the development of smart cities, the algorhythmic battlespace follows what Orit Halpern and Gökçe Günel have theorized as the "particular form of spatial and temporal containment and speculation"[151] engendered by a logic of experimentation and prototyping, a form of politics of "qualculation"[152] that runs counter to judgement, careful deliberation, and democratic decision-making. By moving toward immersive experiences and real-time actions, reflexivity and planning are eliminated in the name of immediacy, instantly turning indicators into actions without ends,[153] with the purpose harnessing the potentials of emergence.

Here, we can see that the target and the best courses of action emerge together through the process of targeting within the cognitive assemblages—a sort of optimization through action in which the best course is not predefined as per planning methodologies but, rather, emerges through the monitoring/tracking and the operationalization of the feedback loop between affect and learning. This is the military in hyperactive mode, in which "permanent vigilance, activity, and intervention"[154] are conjoined with an experimental logic to (re)design the world according to targeting.

However, this is a practice that is arguably less a product of a specific governmental rationality than a product of the logics inherent in the targeting methodologies and the computational data logics through which it actualizes its operations. The conjoining of military design and data logics thus creates its own rationalities and modes of operating, endlessly translating and remaking the world through violent trial and error and meaning-making operations.

As such, targeting not only precludes other types of representation[155] but is increasingly building and imposing a world through digitalization, datafication, and computing, made for and by targeting, in order to offer the best conditions possible for the functioning of targeting. It is thus not simply a question of an ecology of operations trapped in feedback loops, as with F3EAD in the previous chapter, but rather, that this ecology has been designed and engineered to feed off and transform through violent immersive feedback.

While agency and functionality are messily distributed across the cognitive assemblage of humans, machines, and the environment, so is their dysfunctionality or capacity for failure. These assemblages are not designed as final but are engineered to contingently unfold from a built-in logic of evolution through error, repair, and optimization. What unfolds within these martial algorhythms is then a reordering of time itself that can no longer be thought of as evolving in a specific order or sequence of action. The result is that constant refigurations of the world replaces the teleological direction of ending. Unlike cybernetics and the closed world of Cold War computing, contemporary targeting configurations are not goal-oriented, steering their feedback toward a central struggle, an end, or normative future. Rather, they produce a form of instrumentality that lacks a teleology beyond the operational logic of keeping the process going, in which constant learning through experimentation rather than envisioned end points is the guiding rationale.

War in the Age of Planetary Targeting

Although traditional forms of military targeting operate alongside and in concert with the contemporary targeting assemblage, the convergence of experimental computational logics, and military design thinking, points toward a vision of the future in which military targeting reduces warfare to endless experiments to develop insights into emerging futures. These "experiments in participation" seem like they are straight out of a science and technology studies (STS) handbook, where targeting operates as both a tool of inquiry and a device employed to create and perform collectives.[156]

Military design thinking and the computational regime described above are not problem-solving in the sense that they create a form of insecurity or objects for knowledge and action, but they take the form of a meta-strategy, a whole new way of thinking about problems in a world that is seemingly complex and in continuous emergence, a world in which there are no ends. It evokes not a normative future but, rather, a strictly empirical normal that seeks to become-with an everchanging fluid world. This targeting assemblage thus has much in common with David Chandler's notion of *hacking* "as a set of techniques not really 'making' something but rather acting as a stimulus, exploring, probing, facilitating, repurposing what already exists but can only come into being 'with,'"[157] which sees unknowability as not a problem "but an asset, an invitation to limitless possibilities."[158] In its most radical form this is a martial vision in in which failures are seen as opportunities, enemies are never fully represented or specified but dealt with in the same manner, and constant violent experimentation becomes the new imaginary of a world in the making.

Embracing an experimental ethos of emergent knowing, or knowledge-as-learning, contemporary reconfigurations of military targeting do not make a claim to know in the sense of being a technocratic vision of the world to be made. It is not necessarily about clear-sightedness or arriving at knowledge, a solution, but about devising or putting things together in a way that allows assumptions to find a form, to become tangible according to targeting's martial logics. And once an object of sensation becomes tangible, new contours can be drawn onto the map, making it possible to move deeper and deeper into the terrain. This traversing of scales and the unknown, through design, experimentation, and trial and error works as a form of learning—establishing unstable and fluctuating bearing points functioning as a compass for understanding or a sense of direction.

Learning is thus not about making conclusions or finding clear-cut answers but about designed explorations that enable militaries to place experiences, memories, events, people, and things in a certain order, creating new horizons, (re)constructing operational environments, and enabling new forms of violence. Here, the enemy appears as complex forms of relations, a fluid becoming, always in figuration, algorhythmically modulated through ongoing reconfigurations of humans, technologies, and the environment. The enemy does not need to be fixed but, rather, operationalized as always becoming, continuously probed. This algorhythmic operational environment not only spatiotemporally expand warfare everywhere[159] and *ad infinitum*[160] but make futures compatible, accessible, and user-friendly to military violence, while military operations constantly feed the remodulations of possible future horizons.

With the MoW, targeting is now moving away from seeking to control and direct systems and toward specific desired outcomes as with "destructive functionalism." And it is not simply a way to enable the adaptability and fungibility of an unpredictable unfolding of events in order to prevent, preempt, or manage undesired effects. In many ways, targeting's operations are moving away from being an instrumental transformative force, on par with the famous Clausewitzian dictum of war as an instrument of politics, and toward a supposedly creative and experimental force seeking to experimentally engage with and harness the potentials of emergent processes. Not only do data analytics and so-called artificial intelligence promise novel ways of producing foresight but, together with military design thinking, they are part of a shift in anticipatory thinking and practices that engenders, and in many ways demands, a more proactive violent experimental approach to the world in which violence generates data, data generates learning, and learning generates new violent operations.

Contemporary reconfigurations of military targeting are thus not only "potentially and tendentially lethal [making] the act of targeting ... an act of violence even before any shot is fired"[161] but a martial practice in which warfare is reconceptualized as endless processes of complex experimental interactions and continuous disruptions. It is a vision of war that does not promise victory; it is not seeking to control and establish order but, rather, to dominate emergence, to capitalize on the volatilities of the world, and "track and whack" whatever pops up. This is a world in which it becomes impossible to envision alternatives to violent militaristic thought and practice.

Continuously experimenting with future beginnings through ongoing modulation of an empirical normal, produced by the folding and unfolding of patterns and anomalies in which everything and everyone, everywhere and all the time, has already been targeted, this real-time vison epitomizes a specific spatiotemporal logic that often reduces possibilities to a single horizon—an outlook in which warfare is not only normalized but becomes the only viable, rational, and logical way of engaging with the world. Rather than accepting the Clausewitzian fog and friction as an ontological condition of war with inherent limitations for war as an instrument of policy, this computational targeting regime is increasingly making war possible, unshackling it from its limitations, and releasing its potential[162] through a particular epistemological conviction in which action is seen as learning. The result is an experimental vision of trial and error designed to allow warfare to order the world. This is a world that is increasingly made for military targeting, by military targeting.

7
Conclusions, or Future Beginnings

From the knowns to the unknowns, the static infrastructural targets of the "enemy as a system" and the sociopolitical environment, individuals and things, subjects and objects, signatures and signals, patterns and anomalies, and rhythms and flows, to memories and potential futures—these pages have explored the conditions of possibility for how we ended up in a world where everyone and everything has already been targeted and is continuously retargeted. By highlighting distinct historical glimpses into different worlds of targeting, while at the same time showing how the different historical targeting configurations and their associated techniques have all been brought together within the contemporary algorhythmic present and its accompanying Map of the World, a particular historical problematization of the present has been outlined, setting in motion a specific vision of history and a critical research practice that, hopefully, have offered exploratory fragments into how we ended up where we are.

While I have sought to inscribe military targeting through the genealogies of statistical thought and practice and the historical transformation in epistemic practices and techniques of translation in relation to machines, and to make a claim for a deeper engagement with the technical and functional elements of martial practices and processes, and particularly, the martial apparatus's epistemic operations, I have at the same time argued for a contextual and sociohistorical understanding of these operations and their workings. As such, it has been important to show that the distinct figures of the enemy are not merely political discursive constructs, sensory perceptions, memories, mathematical certainties, aesthetic judgments, or moral convictions. Rather, they emerge as relational effects of historically specific configurations of imaginaries, humans, machines, and environments, and the epistemic operations these engender. These analytical lenses have invited a reassessment of the ways in which we understand military knowledge production and the transformative, generative, and emergent character of war.

This has been illustrated by focusing on how the processes of military targeting—traversing the knowns and unknowns of the world with different techniques of capture—have throughout its history altered the ways in which the martial apparatus understands, makes sense of, and assigns meaning, logic, and rationality to the problem of war and the enemy, and how these different abstracted understandings have violently interacted with the worlds that are made up. In this way, warfare is not only a result of the presence or the actions of a discursively produced enemy but is also engendered by the martial assemblage producing and operationalizing the enemy. Whether through the production of the enemy as a system (operationalized through systems warfare), the enemy as signatures and the transformations toward manhunts, or the enemy as anomalies in spatiotemporal continuums driving military operations towards affective learning and future beginnings, military targeting and its operations have been central to the conditions of possibility for warfare and the ways in which war is imagined, assembled, and waged. The result is a critical narrative drawing attention to how military targeting works as a meeting point between multiple genealogies and different disciplines, highlighting how particular ecologies of operation are contextually entangled with specific politics, histories and imaginaries, and questions of dominance, power, and knowledge.

One of the aims of this investigation has been to open up the question of what war and violence are, inviting us to see beyond damage narratives and extend our analyses of violence into further spaces and temporalities. While this may at times have led the book adrift into the abstract world of military targeting and its many complex and messy operations, this was not meant to dismiss the importance of reporting on the unfathomable number of victims of military targeting or bearing witness to the horrors of destruction. On the contrary, what is ultimately at stake here is the continuous and unevenly distributed and horrific destruction of individuals, societies, populations, and cities. However, the book has been an attempt to address what I see as one, or more accurately, a series, of the most important and urgent issues confronting us today, with implications that extend far beyond the immediate horrors on the ground. There is a pressing need to understand the complexities and sociotechnologies of military targeting and its world-making practices that are violently reordering our world, weaving war deeper and deeper into the fabrics of everyday life and the connective tissues of our planetary lifeworlds, in which war and violence are seen as the default option, not the exception, for politics.

Future Beginnings and the Problem of Critique

The contemporary targeting ecology, in the form of the Map of the World, is not only a database through which the enemy is modeled, military spacetime is constructed, and war is transformed. It is an ecology transforming how the world can be known, how it is perceived, and how it is operationalized. Earlier databases, such as the Bombing Encyclopedia, the Hamlet Evaluation System (HES), and kill lists of the Phoenix program and the war on terror all produced and archived wholes out of the particular—systematically and relationally linking targets to make up the model of the enemy. These simultaneously worked as active operational sites for epistemic processes and planning with an ability to recombine and remodel the enemy as new data entered the systems.

Although many have argued that digital computation changes the nature of the archive,[1] this book has shown that databases are no longer merely containers of possibilities but are environments or ecologies, where the world is not just translated, made known, and perceived but is experienced. With the Map of the World and the quest to immerse the martial apparatus in data, all things are brought into relation in specific ways, generating new forms of bioconvergence as epistemic operations constantly make and remake worlds.

Understanding the world as complex and in a state of becoming, carefully calculated effects, based on specific means and ways, have given way to affects in which knowledge and understanding have been supplanted by learning through doing and experience. This martial epistemology opens up toward the unexpected, the unplanned, and the unprogrammed, where the possible is always bound up with affective capacities of experience and the productive capacities of error and fallibility. With the world constantly being refigured and remodeled, we are now entering an era in which martial algorhythmics, coupled with older targeting techniques, are framing the world as an experimental operational environment seeking to harness potential futures from data. As such, the Map of the World has become a self-referencing ecosystem in a permanent state of emergence, ready to activate kinetic and nonkinetic violence in a world where everything and everyone is a target of opportunity, and everything is available for weaponization.

This digital twin promises not only to transform what kinds of data will be ingested, processed, and stored—where, how, and by what and whom— and what can be discovered and analyzed about the world, but declares itself as the horizon of our possible worlds and its futures as it translates the world into data and the data into worlds. This omniscient dream is, of course, grossly exaggerated—not only because most of what goes on in war is still not datafied, processed, analyzed, and databased but because, as George Box's famous dictum says, "All models are wrong."[2]

While this saying was once expressed in relation to the limits of statistics, and is concurrent with the views of Breiman discussed earlier, it is also valid in terms of the limits of machine learning, algorhythmics, and so-called artificial intelligence. Flattening, processing, and modeling the world in particular ways, through imaginaries, configurations, and epistemic operations will always portray a situated, abstracted, and selected version of the world. However, this is only part of the problem. The rest of the famous dictum tell us that while all models are wrong, "some are useful."[3] But, for whom are these models useful?

Though the sensors, data processing technologies, and machine learning techniques discussed in these pages have, in large part, supplanted the traditional epistemological conviction based on causation and explanation in favor of automated correlations, their perceived usefulness for the martial apparatus is clear. In the minds of the martial apparatus, this ecosystem and its spatiotemporal processes are working, because it will always, by its very epistemic configurations, churn out and find causal links, correlations, patterns, signatures, anomalies, and possibilities. And it is this usefulness, understood as the generation of different actionable hypotheses and possibilities, that is now utilized as a way to access the world, seeking to merge humans with the datascape of the Map of the World to activate a violent becoming-with.

In addition, as we have seen, because algorhythmic targeting is based on learning through action, which again, is determined by spatiotemporal movement, these operations depend on volatility in the environment. Each violent action generates new data, and new data are what, in this technoscientific imaginary, enables learning and new possibilities for action. The logic here is that by inducing volatility through the capacities of the configurations one can generate interactions of a different kind and "cause" data to appear in order to support the cyclical nature of the system. Without volatility, the systems grind to a halt. This creates another imperative to act and an operational logic that is not based on effects, computed as step-by-step processes to achieve a certain outcome, be it the collapse of a modern industrial state or a network, but is based on a logic that demands constant probing operations to learn.

The Map of the World and its accompanying techniques are thus not about surveillance or the dream of transparency but, rather, about the capacity to immerse the martial assemblage in the world, experience it, and act in it. Its goal is not to understand why something is happening but how it is happening, not to increase knowledge or to understand its meanings, only to make the world actionable and targetable.

While the solution is to act because action generates problems to solve, these are not simply based on recursive goal-directed feedback loops offering single solutions or reducing possible courses of actions. Rather, they are configured so as to generate hypotheses about the future, making the future a flexible frontier with many possible horizons. The martial apparatus is interested in "phenomena-in-their-becoming"—not stopping them in their becoming—to harness their potential future beginnings and seek constant opportunity. This circular logic thus establishes a military necessity to experiment with the world through data processing and military operations, transforming martial epistemology in ways that pose particular problems for critique.

Machine Reasoning, Disruptive Design, and the Destruction of Ends

The seemingly reductionist logics and empiricist epistemologies of computation and design can be critiqued in traditional ways, exposing the material and discursive infrastructures that sustain these logics and practices, the organization of violence and politics that this regime seems to engender or even foreclose, or pointing out the inherent contradictions, computational opacities, biases, hubris, and fault lines that targeting rests on. But the contemporary condition also calls for moving beyond the traditional critique of uncovering hidden truths, or how to think from within particular techniques, tools, and technologies. The problem for critique lies in how to engage with what appears to be a break in rationality and logic as operational design and artificial forms of reasoning are increasingly merging, where deductive, inductive, and abductive reasoning intermingle.

What happens to critique when computational technoscience and military design do not make an empirical claim to "know," in the sense of being a technocratic vision of the world to be made? What happens to the object of critique when they reject the inherent reductionism of modernism, which has been a core object of critical inquiry, in favor of an open, all-encompassing view of the world that can be accessed through constant probing and experimentations and experienced through immersion fueled by advances in computational sciences? When the unknown is seen as a potential, operations have become learning, the immersive sensorium is knowledge, failure is opportunity, ends are redefined as constant experimental engagements with future beginnings, war is unhinged from its supposedly instrumental political goals, and cognitive assemblages are seemingly operating in opaque ways trading interpretability for accuracy, how do we account for this? The inflection points for critique, I argue, are in the details of the interactions between humans, machines, and the environment in ecologies of operations.

What is at stake with the development of machine learning and so-called artificial intelligences is that the contemporary targeting regime is not merely about the automation of perception, memory and knowledge production in terms of an efficient way of executing specific tasks based on preestablished rules. Today's neural net models "learn" by extracting patterns from vast datasets without following explicit programming, establishing their own so-called ground-truths or sets of assumptions. They are thus the inverse of rule-bound manipulable mobiles, not deducing from an output of a preprogrammed algorithm but, rather, seeking to find an optimal model that produces this output.[4] If algorithms and algorhythmics are no longer rule-bound logical series consisting of a number of steps but, rather, generate output that is beyond input rules in an abductive way, they can therefore not be thought of, analyzed, or critiqued in ordinary ways as the instrumentalization of knowledge.

Prior to so-called generative artificial intelligence and deep neural networks, assumptions and meaning were inscribed into configurations at all levels through design and engineering practices as a set of rules. These configurations also involved multiple interactions between humans and machines, from the selection and labeling of data to the adjustment of algorithms based on the output, to the interpretation of data. In all these instances, imaginaries would play a role to instill meaning and instrumentality to the processes. This would change if the systems were only working on raw data through which the algorithms and deep neural networks were able to modify themselves in and through their encounters with environmental data.[5] These would not need the set of assumptions, or imaginaries, that provide rule-bound steps to create meaning. This, however, does not mean that artificial intelligence is unknowable or operates in mysterious ways, as is often argued, but means that we must understand the context in which it operates and its end-to-end operations.

Although this book has only skimmed the peripheries of the micropractices of algorhythmic targeting and the Map of the World, one can only assume that meaning, in the cognitive assemblage sense, is added at the output level. We know that ABI does not work from a predefined notion of what is either normal or anomalous and that the selection or labeling of data is done without a specific notion of what matters, but that this is learned through ABI's systemic encounters with data. However, we also know that the anomalies produced by ABI, which trigger an alert in the system that is checked by a human analyst, are either fed back into the system as normal and then become part of the empirical baseline or are deemed actionable or investigative objects. Therefore, we can speculate that meaning is in large part installed at the output level through interpretation. This does, of course, have consequences, not least because it is increasingly unclear what these systems learn and thus what these outputs actually are.[6]

What is clear, however, is that this is not leading to a reduction of the so-called fog-of-war. Rather, it creates a "machinic" haze, reproducing the self-referencing and closed world of military targeting. This is because in order to make these systems operate in the openness and contingency of the real world, it is necessary to design the world as closed. In other words, making the environment computable and the computer environmental at the same time as making warfare compatible with the world, it is necessary to close the world to the extent that the "incomputable" does not hamper its functions. But to complicate matters, this closed world does not refer back to an end state or a central struggle, as in Edwards's conclusions,[7] but is geared toward generating endless future beginnings.

These algorhythmic systems and processes are part of an ongoing transformation of the conditions of thinking, reason, and logic. If human perspectives are now recursively entangled with what algorithmic models do and know, learning and understanding acquire a new meaning. The question then concerns not only the cognition of programmatic functions or the execution of rules but how the experimental ethos of internalizing errors, fallibility, and unknowns and constructing hypothetical alternative scenarios works and with what consequences. These techniques do not simply reveal an unknown world to us that we have tasked the automated systems to find, as is the common rhetoric of designers and decision-makers, but they create new worlds. Violent experimental worlds. Worlds that are at the same time a different reduction of reality but nevertheless engender new possibilities for making it actionable.

Working from an epistemology that has internalized failures and errors as part of its operational logic of learning, the contemporary configuration of targeting has created a novel form of spatiotemporality and a new form of instrumentality that is radically different from the teleological visions of the past or forms of technical governance associated with geopolitics, technopolitics, biopolitics, or the society of control.

It does not have distinct reference objects, normative end-states, or goals to strive for or manage, either through contingency planning or pre-emptive operations to thwart dangers in their making. The contemporary world of targeting only caters to the functioning of targeting, demanding more and more sensors, more and more data, and increased automation of epistemic operations.

Conclusions, or Future Beginnings

Although these systems form a core part of the contemporary targeting configuration and the Map of the World, it would be unwise to wholly dismiss the sociopolitical aspects of this targeting regime vis-à-vis its technical functions and operationality. While contemporary targeting poses as an empiricist model, it stems from sets of assumptions or imaginaries that are crucial to understand. Situated practices and context matter for how the immersed analysts experience the Map of the World and for the way in which the "algorithms are immersed in the knowledge that the analysts have in their minds about intelligence problems."[8]

(Re)Configuring Critique

Configurations and cognitive assemblages, as this book has shown, offer a way to study the technoscientific and sociopolitical together, emphasizing how military imaginaries are designed and engineered into systems and how both are (re)configured through its encounters with the world. While the details of how exactly these systems operate remain elusive, configurations have offered a methodological lens through which both military imaginaries and the computational modalities that animate these systems can be assessed. More detailed work digging into the mechanisms of martial algorhythmics certainly needs to be conducted, but this book has provided important nuances of the interrogation of war and warfare and, by doing so, has aided in framing different, but important, critical questions about contemporary war and warfare. Going forward, however, it is important that we ask the right questions.

The introduction of so-called artificial or synthetic intelligences, machine learning techniques, and autonomous systems into warfighting assemblages forces us to reflect on what thinking, reasoning, and knowledge and war are becoming through these technical elements. If one takes seriously Hayles's argument that computers are cognizers—that they produce and process meanings, for humans, for themselves, and for other devices—we must explore not only how the imbrications of humans and machines lead to different kinds of actions, decision-making, and thinking but also, and perhaps more importantly, new forms of choice-making. Then, if decision-making is a joint effort between humans and machines, and part of a broader relationally entangled mesh of prior choices, judgments, and decisions, this is were politics is located. One could, of course, claim that to make a problem a part of a computational process, or even to build a computing system, is a human choice made without machines, but this would neglect the way in which computing, and its promises, may reside in our imaginations and desires, with the capacity to form and affect human choices. Such claims also tend to overlook that computation and other technical practices often exceed human choice or intention and can thus not simply be seen as an automation of human instrumentalization. By way of Hayles, computers and algorithms are not only mathematical calculative tools employed in the service of humans but are true cognitive entities that can create meaning through a mode of reasoning and data processing that is beyond human cognition and is in many ways alien to humans. As such, they differ from previous technologies that humans have employed for identifying and sorting, for classifying, categorizing, and indexing in the name of security and war.

Classification practices are intrinsically political,[9] but today, algorithmic models are increasingly making these classifications, working through masses of data, creating similarities and differences, for choices to be made. Thus, agency and power become distributed in messy ways—a point that is especially significant as complex human-technical systems and materialized cognition become increasingly ubiquitous and humans are only one kind of cognizer among many. Therefore, part of understanding machines is the need to "extend cognition, intention and meaning to nonhuman subjects and computational networks."[10] If machines are now generating outputs and modifying themselves on their own through their encounters with data, we also must ask ourselves different questions than previously, questions that ask not only the "what if" of future technologies, but the "what is," "what they do," and "how they do it."

What happens when questions about war and warfare not only become questions of calculation but of machine learning? How do machines think and cognize? What is it that machines learn? What is the "human" in these systems? How do different cognizers experience war? Who or what has agency? Where is choice located, and are decisions now operating as "already-performed" due to the multitude of choices already made? Only through such questions can we begin to rethink the ways technologies and the epistemological transformation in thinking in relation to machines have changed the way we see the world and transformed the relationship between sensing and making sense, between knowing and doing, understanding, and acting.

The immediate answers to these questions are that context matters and that they are empirical conundrums that cannot be answered a priori without detailed explorations of the configurations and epistemic operations that make these possible. If we want to engage with the material and discursive infrastructures that hold the logics of military targeting together and make certain practices and effects possible, importance must be placed on the ways in which configurations are both designed and work, that is, their

Conclusions, or Future Beginnings

217

end-to-end operations—how they produce data and with what kinds of instruments, how data are processed and analyzed and with what kinds of techniques, and how the resultant outputs are interpreted and operationalized.

Here, historical investigations of the conditions that make it possible for these war-fighting assemblages to emerge also have a role to play. Not only can they question the novelty of many of the claims regarding contemporary sociotechnical systems, but they can also reveal how we got here, problematizing the present and pointing out the many ways in which these systems are dependent on historical imaginaries, assumptions, knowledges, and ideas about power and domination, politics, and war. If something is invented, there is also a possibility to "uninvent" it as a rational and legitimate solution to a defined problem or as the solution that defines the problem. This does not mean reverse-engineering these systems but, rather, pointing out their historically contingent making, the worlds these configurations are making and the worlds they are excluding.[11] Thus, aiming at a diagnostic screening of the present through a constructive engagement with the past can open up novel avenues for critically analyzing pressing social issues.

While these questions might seem like they are upholding the binary between humans and machines, they must be asked at the level of configurations, cognitive assemblages, and ecologies of operations. Following Suchman and Hayles, the inflection points for critique lie not in the intrinsic capacities of either human or machine but in the capacity for action that arises out of configurations and ecologies of operation. In the same way that humans are not autonomous from the things and relations that surround them, computers and machine learning techniques are not simply processing devices devoid of human context, interpretation, and meaning-making. For instance, the available data affect and condition the epistemic operations whether these are humans, machines, or a combination.

For all the talk about artificial intelligence, there is a need go beyond the notion that intelligence is linked just to the brain or the so-called neural network and recognize that both humans and machines also learn though their senses and sensors, shaping their experiences of the world, and that they do so together. And it is here, in the relations and interactions between humans and machines and between the sociotechnical systems and the environment, that critique can find its way. This means that it is the complex agencies within the assemblages that must be analyzed, and the different capacities that arise from these. Thus, we must find ways of critiquing not only how we, as humans, design and engineer our societies but also what assemblages do, how they experience the world and wars, how and what they "think," and importantly, how they figure us.

By foregrounding the present as a question of configurations and cognitive assemblage, we can move beyond pointing out the opacity, biases, and fallibility of novel computational systems, critiquing their claims of empiricism and deductive logics and the inherent trading of interpretation for accuracy in algorithmic modeling cultures. This means that critique must acknowledge that the problem is neither humans nor machines and accept that agency is messy and that both humans and machines now think and increasingly do so together in cognitive assemblages. Through such sensitivities one can ask and critique how worlds are produced, how choices are generated, where decisions lie, and importantly, how meaning is given in data.

If military targeting and its accompanying algorhythmic processes have now cast aside war as an instrument of politics towards desired end states, there is also a need to rethink instrumentality, not only functionality—to move beyond exploring and critiquing what means do or how means have taken over the ends and move toward thinking about the logic of future beginnings in cognitive assemblages and examine whose instrumentality is being mediated and operationalized. This must be based not on a new technopolitics that only states the ontogenetic limits of technoscience—or war, for that matter—and the internal limits of algorithmic programming but one that engages deeply with the experimental logics of machines and military design and the move beyond ends. However, this also requires us to take war seriously, as a processual phenomenon in and of itself that is not first and foremost a political instrument derivative of social and political orders but is, in fact, also constitutive of and reshaping these orders in important ways.

Doing so may enable a reopening of the political dimension of instrumentalism to include the automation of logical thinking that is part of a new mode of reasoning. Such an approach might offer ways to reimagine a technopolitics that is not teleological or instrumental but that nonetheless has politics. That is, a politics that turns away from the dogmatic "real-time" military design model of algorhythmics transforming the individual, the social, and the material into a laboratory of trial and error in which warfare is based on the functional informational feedback loops and hypothesis-generation afforded by the affective capacities of the targeting configuration, always present, slow but at the same time abrupt, seeking out emergent opportunities and ready to strike. This may be a lofty hope, but it certainly starts with accepting that humans and algorhythms together produce and act upon the world.

This may also help to move many of the debates on automation and autonomy away from an entrenched perceived binary of humans either outside or inside the loop, which is at the core of, for instance, the lethal

Conclusions, or Future Beginnings

autonomous weapons debate, and toward consideration of new forms of agency and decision that are now distributed across messy assemblages of sociotechnically produced battlespaces and targets. Acknowledging that humans and machines are already intertwined within the loop, that there is no "autonomy" of either, so to speak, and acknowledging that machines, and not only humans, think and act, will be crucial for imagining an alternative future of politics. It also means that we must accept that decisions and actions are derived from a long process of human-machine-environment interactions that involves a number of choices along the way—from the design and engineering of configurations to the myriad of interactions and relations these generate, to the production and processing of data and interpretations of outputs all the way to decisions and actions.

This is not least important as the US war-fighting machinery has newfound optimism regarding the dream of network-centric warfare (NCW) and so-called revolutions in military affairs (RMA). Armed with the promises of artificial intelligence, the US martial apparatus is currently embarking on a massive endeavor to connect every sensor, platform, operator, and effector into a giant "internet of military things" under the moniker of All-Domain Command and Control (JADC2),[12] accompanied by operational concepts to take this network to the frontlines in the form of Mosaic Warfare.[13] Turning its back on massive-scale manhunts for the foreseeable future and refocusing its martial attention on China, systems thinking and systems warfare is back in vogue, updated with autonomous sensors and robotic swarms, all operating on and through the immersive capacities of the Map of the World, constantly updating the real-time situational awareness ecosystem. In this imaginary future conflict with China, artificial intelligence and autonomous weapons systems are no longer seen as an option but have become an imperative. Configuring imaginaries about war into artificial intelligences and imaginaries about artificial intelligences into massive war-fighting machineries, this is a world in which the only rational and logical way of thinking about world politics, international relations, and security is through military supremacy, global domination, and technological preeminence, closing off alternatives other than the steady march toward confrontation.

"The continual targeting of the world as the fundamental form of knowledge production," Rey Chow argued in relation to strategic bombing and nuclear weapons, "is the inability to handle the otherness of the other beyond the orbit that is the bomber's own visual path."[14] Today we can expand on this insightful observation by arguing that it is not only the Other who is made up by targeting. Everyone and everything are always targeted. Targeting is ultimately also how we come to see and understand ourselves and our own societies—from the mirror imaging after WWII, to cybersecurity, to critical infrastructure protection, to so-called information operations, and to the fears of artificial intelligence today. How we view ourselves through the epistemologies of targeting ultimately has consequences for how we view others and our relationship to otherness. In this way, to channel Karen Barad again, the practices of knowing and being and taking action are mutually implicated. We now live in a world in which war and knowledge feed off and enable each other in ways that foster the emergence of a global order where targeting is the protagonist.

There will always be trade-offs, but these can only be understood if we start thinking seriously about what it means to live in a world that is forged in the imaginaries, operative logics, epistemologies, and socio-technical practices of military targeting. This is where critical scholars can extend critique to do work—by conveying these problems, dilemmas, and fundamentally problematic aspects of what we are creating and showing how there are always different ways of encountering and creating worlds by arranging things.

This is not a call for a vision outside of technogenesis or a return to thinking in cartesian terms but, rather, a warning that we, those of us concerned with the current and future trajectories of martial logics and operations, must be aware of what is being created through, with, and from technologies. Technologies are not fundamentally deterministic, but they are part of what determines how configurations produce the world and act in it—reproducing, institutionalizing, doctrinizing, and transforming the military imaginaries of targeting that gave rise to them. Intervening in and contending with contemporary transformations in martial epistemology and (re)configurations of targeting is a task that seems as urgent as ever, as we are all now living in an experimental multidimensional digital twin of the world, made for military targeting by military targeting.

Conclusions, or Future Beginnings

Notes

Chapter 1

1. Michael Vincent Hayden, *Playing to the Edge: American Intelligence in the Age of Terror* (New York: Penguin Press, 2016), 31.

2. Seymour M. Hersh, "Manhunt: The Bush Administration's New Strategy in the War against Terrorism," *New Yorker*, December 15, 2002, 66.

3. Hayden, *Playing to the Edge*, 32.

4. National Security Agency (NSA), "Joint Document Highlights NGA and NSA Collaboration," news release, December 27, 2004.

5. Hayden, *Playing to the Edge*, 136.

6. Ian Hacking, "Making Up People," in *Reconstructing Individualism*, ed. Thomas C. Heller and Christine Brooke-Rose (Stanford, CA: Stanford University Press, 1986).

7. See, for instance, Antoine Bousquet, *The Eye of War: Military Perception from the Telescope to the Drone* (Minneapolis: University of Minnesota Press, 2018).

8. Paul N. Edwards, *The Closed World: Computers and the Politics of Discourse in Cold War America* (Cambridge, MA: MIT Press, 1996).

9. Peter Galison, "War against the Center," *Grey Room* 1, no. 4 (2001): 29

10. Letitia A. Long, "Activity Based Intelligence: Understanding the Unknown," *Intelligencer: Journal of U.S. Intelligence Studies* 20, no. 2 (2013): 7.

11. US Department of Defense, "DoD News Briefing—Secretary of Defense Donald H Rumsfeld and General Richard Myers, Chairman, Joint Chiefs of Staff, News Transcript," news release, February 12, 2002.

12. The figure is taken from Long, "Activity Based Intelligence," 9. Available at https://www.afio.com/publications/LONG_Tish_in_AFIO_INTEL_FALLWINTER2013 _Vol20_No2.pdf.

13. The figure is taken from Government Communications Head Quarters (GCHQ), *GCHQ Analytic Cloud Challenges* (2012), 17. https://assets.aclu.org/live/uploads /document/foia/GCHQAnalyticCloudChallenges.pdf

14. Claudia Aradau and Tobias Blanke also note the relationship between Rumsfeld's knowns and unknowns and the GCHQ matrix (see figure 1.1) but focus on the unknown unknowns and the analytical challenge of "discovery," or anomaly detection. See

Claudia Aradau and Tobias Blanke, "Governing Others: Anomaly and the Algorithmic Subject of Security," *European Journal of International Security* 3, no. 1 (2018): 1–21.

15. Anna Mulrine, "Warheads on Foreheads," *Air Force Magazine* 91, no. 10 (October 2008): 44–48.

16. Ashley M. Richter, Rupal Mehta, and Michael Hess, "The Frontier of Multimodal Mapping: The Future of Secure, Integrated Data Visualization," *Trajectory Magazine*, January 25, 2019. https://medium.com/the-2019-state-and-future-of-geoint-report/the-frontier-of-multimodal-mapping-8247c27038cb

17. Michel Foucault, *Discipline and Punish: The Birth of the Prison* (New York: Penguin, 1977), 31.

18. L. D. Kritzman, "Power and Sex: An Interview with Michel Foucault," in *Michel Foucault: Politics, Philosophy, Culture: Interviews and Other Writings, 1977–1984*, ed. L. D. Kritzman (London: Routledge, 1988), 262.

19. See, for instance, David Garland for a discussion of the importance of "diagnosis" for this kind of genealogical approach. David Garland, "What Is a "History of the Present"? On Foucault's Genealogies and Their Critical Preconditions," *Punishment & Society* 16, no. 4 (2014).

20. Stuart Elden, *Mapping the Present: Heidegger, Foucault and the Project of a Spatial History* (London: Continuum, 2002).

21. Stuart Elden, "Land, Terrain, Territory," *Progress in Human Geography* 34, no. 6 (2010): 2.

22. This approach is inspired by, Jussi Parikka, *Operational Images: From the Visual to the Invisual* (Minneapolis, USA: University of Minnesota Press, 2023).

23. Kevin McSorley, "Archives of Enmity and Martial Epistemology," in *(W)Archives: Archival Imaginaries, War, and Contemporary Art*, ed. Daniela Agostinho et al. (Berlin: Sternberg Press, 2021).

24. See, for instance, Alison Howell, "Forget 'Militarization': Race, Disability and the 'Martial Politics' of the Police and of the University," *International Feminist Journal of Politics* 20, no. 2 (2018): 117–136.

25. On this conceptualization of epistemology, see, Hans-Jörg Rheinberger, *On Historicizing Epistemology: An Essay* (Stanford, CA: Stanford University Press, 2010).

26. Karin Knorr Cetina, *Epistemic Cultures: How the Sciences Make Knowledge* (Cambridge, MA: Harvard University Press, 1999).

27. Rheinberger, *On Historicizing Epistemology*, 2.

28. See, for instance, Ian Hacking, *Historical Ontology* (Cambridge, MA: Harvard University Press, 2002); Lorraine Daston, "Historical Epistemology," in *Questions of Evidence*, ed. James K. Chandler, Arnold Ira Davidson, and Harry D. Harootunian (Chicago: University of Chicago Press, 1994); Lorraine Daston and Peter Galison, *Objectivity* (New York: Zone Books, 2007).

29. Karen Michelle Barad, *Meeting the Universe Halfway: Quantum Physics and the Entanglement of Matter and Meaning* (Durham, NC: Duke University Press, 2007);

Donna Haraway, "Situated Knowledges: The Science Question in Feminism and the Privilege of Partial Perspective," *Feminist Studies* 14, no. 3 (1988): 575–599.

30. Lucy Suchman, *Human-Machine Reconfigurations: Plans and Situated Actions*, 2nd ed. (Cambridge, UK: Cambridge University Press, 2007); Derek Gregory, "Gabriel's Map: Cartography and Corpography in Modern War," in *Geographies of Knowledge and Power. Knowledge and Space*, ed. Peter Meusburger, Derek Gregory, and Laura Suarsana (Dordrecht, The Netherlands: Springer, 2015); Anna Danielsson, "Producing the Military Urban(s): Interoperability, Space-Making, and Epistemic Distinctions between Military Services in Urban Operations," *Political Geography* 97 (2022), 102649.

31. Barad, *Meeting the Universe Halfway*, 185. This, of course, also includes me, as a researcher. I do not stand outside of the world but am part of it.

32. Lorraine Daston, *Biographies of Scientific Objects* (Chicago: University of Chicago Press, 2000) 3.

33. Lucy Suchman, "Knowing War," *Political Anthropological Research on International Social Sciences* 4, no. 2 (2023): 131–141.

34. Judith Butler, "What Is Critique? An Essay on Foucault's Virtue (2000)," in *The Judith Butler Reader*, ed. Judith Butler and Sara Salih (Malden, MA: Blackwell, 2003).

35. Etienne Balibar, "What's in a War? (Politics as War, War as Politics)," *Ratio Juris* 21, no. 3 (2008): 365–386.

36. Antoine Bousquet, Jairus Grove, and Nisha Shah, "Becoming War: Towards a Martial Empiricism," *Security Dialogue* 51, no. 2–3 (2020): 99–118.

37. Deborah Cowen and Emily Gilbert, eds., *War, Citizenship, Territory* (New York: Routledge, 2007); Jairus Victor Grove, "Ecology as Critical Security Method," *Critical Studies on Security* 2, no. 3 (2014): 366–369

38. Jairus Victor Grove, "War and Militarization," in *The Oxford Handbook of Global Studies*, ed. Mark Juergensmeyer et al. (Oxford, UK: Oxford University Press, 2018).

39. Bousquet, Grove, and Shah, "Becoming War."

40. Dan Öberg, "Ethics, the Military Imaginary, and Practices of War," *Critical Studies on Security* 7, no.3 (2019): 207.

41. Matthew Ford, "The Epistemology of Lethality: Bullets, Knowledge Trajectories, Kinetic Effects," *European Journal of International Security* 5, no. 1 (2020): 77–93.

42. Öberg, "Ethics, the Military Imaginary, and Practices of War."

43. Alison Howell, "Neuroscience and War: Human Enhancement, Soldier Rehabilitation, and the Ethical Limits of Dual-Use Frameworks," *Millennium* 45, no. 2 (2017): 133–150.

44. Astrid H. M. Nordin and Dan Öberg, "Targeting the Ontology of War: From Clausewitz to Baudrillard," *Millennium: Journal of International Studies* 43, no. 2 (2015): 392–401.

45. Mathias Delori and Vron Ware, "The Faces of Enmity in International Relations. An Introduction," *Critical Military Studies* 5, no. 4 (2019): 299–303.

Notes to Chapter 1

46. Tarak Barkawi, "Peoples, Homelands, and Wars? Ethnicity, the Military, and Battle among British Imperial Forces in the War against Japan," *Comparative Studies in Society and History* 46, no. 1 (2004): 134–163.

47. Carol Cohn, "Sex and Death in the Rational World of Defense Intellectuals," *Signs: Journal of Women in Culture and Society* 12, no. 4 (1987); Christophe Wasinski, "On Making War Possible: Soldiers, Strategy, and Military Grand Narrative," *Security Dialogue* 42, no. 1 (2011): 57–76.

48. Peter Galison, "The Ontology of the Enemy: Norbert Wiener and the Cybernetic Vision," *Critical Inquiry* 21, no. 1 (1994): 228–266.

49. Coleen Bell, "Hybrid Warfare and Its Metaphors," *Humanity: An International Journal of Human Rights, Humanitarianism, and Development* 3, no. 2 (2012): 225–247.

50. Aradau and Blanke, "Governing Others."

51. Edward W. Said, *Orientalism* (London: Penguin Books, 2003), 98.

52. David Campbell, *Writing Security: United States Foreign Policy and the Politics of Identity* (Minneapolis: University of Minnesota Press, 1992); Iver B. Neumann, "Self and Other in International Relations," *European Journal of International Relations* 2, no. 2 (1996): 139–174; Jef Huysmans, *The Politics of Insecurity: Fear, Migration and Asylum in the EU, New International Relations* (London: Routledge, 2006); Lene Hansen, *Security as Practice: Discourse Analysis and the Bosnian War* (London: Routledge, 2006).

53. Barry Buzan, Ole Wæver, and Jaap De Wilde, *Security: A New Framework for Analysis* (London: Lynne Rienner Publishers, 1998); Claudia Aradau, Luis Lobo-Guerrero, and Rens van Munster, "Security, Technologies of Risk, and the Political: Guest Editors' Introduction," *Security Dialogue* 39, no. 2–3 (2008): 147–154.

54. Vivienne Jabri, *War and the Transformation of Global Politics* (London: Palgrave, 2007).

55. Michael Dillon and Julian Reid, *The Liberal Way of War: Killing to Make Life Live* (London: Routledge, 2009).

56. Judith Butler, *Frames of War: When Is Life Grievable?* (London: Verso, 2010).

57. See, for instance, Aradau and Blanke, "Governing Others."; Lauren Wilcox, "Embodying Algorithmic War: Gender, Race, and the Posthuman in Drone Warfare." *Security Dialogue* 48, no. 1 (2017): 11–28.

58. Geoffrey C. Bowker and Susan Leigh Star, *Sorting Things Out: Classification and Its Consequences* (Cambridge, MA: MIT Press, 1999).

59. David H. Petraeus, "Multi-National Force-Iraq Commander's Counterinsurgency Guidance," *Military Review* 88, no. 5 (2008): 2.

60. Michael J. Shapiro, *Violent Cartographies: Mapping Cultures of War* (Minneapolis: University of Minnesota Press, 1997); Derek Gregory, *The Colonial Present: Afghanistan, Palestine, Iraq* (Malden, MA: Blackwell, 2004).

61. Galison, "The Ontology of the Enemy," 230–231.

62. See Knorr Cetina on how different scientific objects exist in different forms. Karin Knorr Cetina, "Objectual Practice," in *The Practice Turn in Contemporary Theory*, ed. Theodore R. Schatzki, Karin Knorr Cetina, and Eike von Savigny (London: Routledge, 2001), 182; Lene Hansen has also argued that the Other is situated within a "web of identities" rather than the friend/enemy, self/other dichotomy. Hansen, *Security as Practice*, 36.

63. Hacking, *Historical Ontology*, 1.

64. Craig Jones, *The War Lawyers: The United States, Israel, and Juridical Warfare* (Oxford, UK: Oxford University Press, 2020).

65. Jon Agar, "What Difference Did Computers Make?" *Social Studies of Science* 36, no. 6 (2006).

66. Josef Teboho Ansorge, *Identify and Sort: How Digital Power Changed World Politics* (New York: Oxford University Press, 2016).

67. Butler, Judith. "What Is Critique? An Essay on Foucault's Virtue." Transversal Texts, May 2001. www.https://eipcp.net/transversal/0806/butler/en.

68. Antoine Bousquet, *The Scientific Way of Warfare: Order and Chaos on the Battlefields of Modernity* (London: Hurst & Company, 2009).

69. Leo McCann, "'Killing Is Our Business and Business Is Good': The Evolution of 'War Managerialism' from Body Counts to Counterinsurgency," *Organization* 24, no. 4 (2017): 4991:515.

70. Edwards, *The Closed World*.

71. Hunter Heyck, *Age of System: Understanding the Development of Modern Social Science* (Baltimore, MA: Johns Hopkins University Press, 2015).

Chapter 2

1. Derek Gregory, "Lines of Descent," openDemocracy, November 8, 2011. https://www.opendemocracy.net/en/lines-of-descent/; Peter Adey, Mark Whitehead, and Alison J. Williams, "Introduction: Air-Target: Distance, Reach and the Politics of Verticality," *Theory, Culture & Society* 28, no. 7–8 (2011): 173–187; Mark Neocleous, "Air Power as Police Power," *Environment and Planning D: Society and Space* 31, no. 4 (2013): 578–593.

2. Katharine Hall Kindervater, "The Technological Rationality of the Drone Strike," *Critical Studies on Security* 5, no. 1 (2017): 28–44; Grégoire Chamayou, *A Theory of the Drone* (New York: New Press, 2015); Kyle Grayson, *Cultural Politics of Targeted Killing: On Drones, Counter-Insurgency, and Violence* (London: Routledge, 2016).

3. Maja Zehfuss, "Targeting: Precision and the Production of Ethics," *European Journal of International Relations* 17, no. 3 (2011): 543–566; Nicola Perugini and Neve Gordon, "Distinction and the Ethics of Violence: On the Legal Construction of Liminal Subjects and Spaces," *Antipode* 49, no. 5 (2017): 1385–1405; Samuel Issacharoff and Richard Pildes, "Targeted Warfare: Individuating Enemy Responsibility," *New York University Law Review,* 88, no. 5 (2013): 1521–1598; Craig Jones, *The War*

Lawyers: The United States, Israel, and Juridical Warfare (Oxford, UK: Oxford University Press, 2020).

4. Paul Alphons et al., eds., *Targeting: The Challenges of Modern Warfare* (The Hague: Asser Press, 2016).

5. But see Samuel Weber, *Targets of Opportunity: On the Militarization of Thinking* (New York: Fordham University Press, 2005).

6. Anna Danielsson and Kristin Ljungkvist, "A Choking(?) Engine of War: Human Agency in Military Targeting Reconsidered," *Review of International Studies* 49, no. 1 (2023): 83–103; Astrid H. M. Nordin and Dan Öberg, "Targeting the Ontology of War: From Clausewitz to Baudrillard," *Millennium: Journal of International Studies* 43, no. 2 (2015).

7. US Department of Defense, Joint Chiefs of Staff, *Joint Publication 3-60, Joint Targeting* (Washington DC, Department of Defense, 2018). See also Nordin and Öberg, "Targeting the Ontology of War," 402.

8. There is a long-standing debate within military theory regarding the so-called operational level of war. While Clausewitz never talked about the operational level of war, only the political, strategic, and tactical levels, an operational level of war was introduced, the story goes, as the increased size of armies in the industrial age and the subsequent expansion of the spatial and temporal dimensions of military operations necessitated the development of an intermediary level or a conceptual bridge between strategy and tactics to connect them in order for human cognition to encompass the increasing complexity of the phenomenon of war. See, for instance, Shimon Naveh, *In Pursuit of Military Excellence: The Evolution of Operational Theory* (London: Routledge, 1997).

9. The Prussian general Carl von Clausewitz famously stated that "war is not merely an act of policy but a true political instrument, a continuation of political intercourse, carried on with other means," thereby cementing the understanding of war as a political tool in Western political and military circles. Carl von Clausewitz, *On War*, trans. and ed. Michael Howard and Peter Paret (Princeton, NJ: Princeton University Press, 1984), 87.

10. US Department of Defense, Joint Chiefs of Staff, *Joint Publication (JP) 3-0 Joint Operations* (Washington DC: Department of Defense, 2011), II-3.

11. US Joint Cheifs Staff, *Joint Publication 3-60*, II-4.

12. US Joint Chiefs Staff, *Joint Publication 3-60*, II-6.

13. US Joint Chiefs Staff, *Joint Publication 3-60*, II-5 (emphasis added).

14. Arturo Rosenblueth and Norbert Wiener, "The Role of Models in Science," *Philosophy of Science* 12, no. 4 (1945): 316.

15. Perugini and Gordon, "Distinction and the Ethics of Violence."

16. Hans-Georg Gadamer, *Truth and Method*, trans. Joel Weinsheimer and Donald G. Marshall, 2nd rev. ed. (London: Continuum, 2004), 291.

17. Rosenblueth and Wiener, "The Role of Models in Science."

18. James C. Scott, *Seeing Like a State: How Certain Schemes to Improve the Human Condition Have Failed* (New Haven, CT: Yale University Press, 1998).

19. Rosenblueth and Wiener, "The Role of Models in Science."

20. Here, we may think of models in the ways in which media studies, following Harun Farocki's notion of "operative images," have thought about images that do not represent objects but are parts of operations. The model of the enemy does not simply represent the enemy but, rather, is part of the operation of targeting and war. Jussi Parikka, *Operational Images: From the Visual to the Invisual* (Minneapolis: University of Minnesota Press, 2023), 40–41.

21. Brian Massumi, *Ontopower* (Durham, NC: Duke University Press, 2015), 15.

22. Massumi, *Ontopower*, 209.

23. Donald A. MacKenzie, *An Engine, Not a Camera: How Financial Models Shape Markets* (Cambridge, MA: MIT Press, 2006).

24. Robert K. Merton, "Three Fragments from a Sociologist's Notebooks: Establishing the Phenomenon, Specified Ignorance, and Strategic Research Materials," *Annual Review of Sociology* 13 (1987): 1–29.

25. Martin Coward, "Networks, Nodes and De-territorialised Battlespace: The Scopic Regime of Rapid Dominance," in *From Above: War, Violence and Verticality*, ed. Peter Adey (New York: Oxford University Press, 2013), footnote 27.

26. Oliver Belcher, "Sensing, Territory, Population: Computation, Embodied Sensors, and Hamlet Control in the Vietnam War," *Security Dialogue* 50, no. 5 (2019): 416–436.

27. Paul Virilio, *War and Cinema: The Logistics of Perception* (London: Verso, 1989).

28. Derek Gregory, "Dis/Ordering the Orient: Scopic Regimes and Modern War," in *Orientalism and War*, ed. Tarak Barkawi and K. Stanski (London: Hurst, 2012); Kathrin Maurer, "Visual Power: The Scopic Regime of Military Drone Operations," *Media, War & Conflict* 10, no. 2 (2017): 141–151; Kyle Grayson and Jocelyn Mawdsley, "Scopic Regimes and the Visual Turn in International Relations: Seeing World Politics through the Drone," *European Journal of International Relations* 25, no. 2 (2019): 431–457.

29. Eyal Weizman, *Hollow Land: Israel's Architecture of Occupation* (London: Verso, 2012).

30. Thomas Hippler, *Governing from the Skies: A Global History of Aerial Bombing* (London: Verso, 2017); Neocleous, "Air Power as Police Power."

31. See, for instance, Caren Kaplan, *Aerial Aftermaths: Wartime from Above* (Durham, NC: Duke University Press, 2018): Katharine Hall Kindervater, "The Emergence of Lethal Surveillance: Watching and Killing in the History of Drone Technology," *Security Dialogue* 47, no. 3 (2016): 223–238.

32. Derek Gregory, "'Doors into Nowhere': Dead Cities and the Natural History of Destruction," in *Cultural Memories*, ed. P. Meusberger, M. Heffernan, and E. Wunde (Heidelberg: Springer Verlag, 2011).

33. Adey, Whitehead, and Williams, "Introduction: Air-Target," 176–177.

34. Stephen Graham, *Cities under Siege: The New Military Urbanism* (London: Verso, 2010).

35. Derek Gregory, "Gabriel's Map: Cartography and Corpography in Modern War," in *Geographies of Knowledge and Power. Knowledge and Space*, ed. Peter Meusburger, Derek Gregory, and Laura Suarsana (Dordrecht, The Netherlands: Springer, 2015).

36. Anna Danielsson, "Producing the Military Urban(s): Interoperability, Space-Making, and Epistemic Distinctions between Military Services in Urban Operations," *Political Geography* 97 (2022): 102649.

37. Scott, *Seeing Like a State.*

38. Belcher, "Sensing, Territory, Population," 416–436.

39. N. Katherine Hayles, "Cognitive Assemblages: Technical Agency and Human Interactions," *Critical Inquiry* 43, no. 1 (2016): 32–55.

40. N. Katherine Hayles, *Unthought: The Power of the Cognitive Nonconscious* (Chicago: University of Chicago Press, 2017).

41. Antoine Bousquet, *The Eye of War: Military Perception from the Telescope to the Drone* (Minneapolis: University of Minnesota Press, 2018.

42. Geoffrey C. Bowker and Susan Leigh Star, *Sorting Things Out: Classification and Its Consequences* (Cambridge, MA: MIT Press, 1999).

43. Sara Ahmed, "Orientations: Toward a Queer Phenomenology," *GLQ: A Journal of Lesbian and Gay Studies* 12, no. 4 (2006): 543–574.

44. Stefan Helmreich, "Reading a Wave Buoy," *Science, Technology, & Human Values* 44, no. 5 (2019): 737–761.

45. Datafication refers to the process of quantifying human and planetary life, rendering things and phenomena in formats that are machine readable, which subsequently can be analyzed and processed by computers. On datafication, see for instance, Ulises A. Mejias and Nick Couldry. "Datafication." *Internet Policy Review* 8, no. 4 (2019).

46. For more on the role of sensors in security practices, see Nina Klimburg-Witjes, Nikolaus Poechhacker, and Geoffrey Bowker, eds., *Sensing In/Securities: Sensors as Transnational Security Infrastructures* (Manchester, UK: Mattering Press, 2020).

47. Leo Breiman, "Statistical Modeling: The Two Cultures," *Statistical Science* 16, no. 3 (2001): 199–231.

48. John Durham Peters, *The Marvelous Clouds: Toward a Philosophy of Elemental Media* (Chicago: University of Chicago Press, 2015), 3.

49. Orit Halpern, *Beautiful Data: A History of Vision and Reason since 1945* (Durham, NC: Duke University Press, 2015), 40.

50. Scott, *Seeing Like a State.*

51. Lucy Suchman, *Human-Machine Reconfigurations: Plans and Situated Actions*, 2nd ed. (Cambridge, UK: Cambridge University Press, 2007).

52. The concept of imaginaries has a long tradition—in political science following Benedict Anderson's book *Imagined Communities* and in STS following George Marcus, Sheila Jasanoff, Lucy Suchman, and others—and has been utilized in large part to analyze collective systems of meaning that enable the interpretation of reality, the setting of priorities, the channeling of funds and investments, and the guiding of simplifications and standardizations. See, Benedict Anderson, *Imagined Communities: Reflections on the Origin and Spread of Nationalism* (London: Verso, 1983); George E. Marcus, *Technoscientific Imaginaries: Conversations, Profiles, and Memoirs* (Chicago: University of Chicago Press, 1995); Sheila Jasanoff, "Future Imperfect: Science, Technology, and the Imaginations of Modernity," in *Dreamscapes of Modernity: Sociotechnical Imaginaries and the Fabrication of Power*, ed. Jasanoff Sheila and Kim Sang-Hyun (Chicago: University of Chicago Press, 2015); Lucy Suchman, "Configuration," in *Inventive Methods: The Happening of the Social*, ed. Celia Lury and Nina Wakeford (London: Routledge, 2012).

53. Michel Foucault, "The Confession of the Flesh," in *Power/Knowledge Selected Interviews and Other Writings*, ed. Colin Gordon (London: Vintage, 1980), 194.

54. Suchman, "Configuration."

55. See, for instance, Martin Coward, "Networks, Nodes and De-territorialised Battlespace: The Scopic Regime of Rapid Dominance," in *From Above: War, Violence and Verticality*, ed. Peter Adey (New York: Oxford University Press, 2013); Dan Öberg, "Ethics, the Military Imaginary, and Practices of War," *Critical Studies on Security* 7, no. 3 (2019): 207; Sean Lawson, "Articulation, Antagonism, and Intercalation in Western Military Imaginaries," *Security Dialogue* 42, no. 1 (2011): 39–56.

56. Parikka, *Operational Images*, 24–25.

57. Bruno Latour, *Reassembling the Social*. (Oxford, UK: Oxford University Press, 2007).

58. Bruno Latour, "Morality and Technology: The End of the Means," *Theory Culture & Society* 19, no. 5–6 (2002): 247–260.

59. Lucy Suchman and Jutta Weber, "Human-Machine Autonomies," in *Autonomous Weapons Systems: Law, Ethics, Policy*, ed. Nehal Bhuta et al. (Cambridge, UK: Cambridge University Press, 2016), 79.

60. Thomas Lemke, "An Alternative Model of Politics? Prospects and Problems of Jane Bennett's Vital Materialism," *Theory, Culture and Society* 35, no. 6 (2018): 31–54.

61. Nordin and Öberg, "Targeting the Ontology of War."

62. Danielsson and Ljungkvist, "A Choking(?) Engine of War."

63. Anna L. Tsing, *The Mushroom at the End of the World: On the Possibility of Life in Capitalist Ruins* (Princeton, NJ: Princeton University Press, 2015), 23.

64. Suchman and Weber, "Human-Machine Autonomies."

65. Jonathan Roberge and Michael Castelle, "Toward an End-to-End Sociology of 21st-Century Machine Learning," in *The Cultural Life of Machine Learning*, ed. Jonathan Roberge and Michael Castelle (London: Palgrave Macmillan, 2021).

66. Tugba Basaran et al., eds., *International Political Sociology: Transversal Lines* (London: Routledge, 2017).

Notes to Chapter 2

67. Grégoire Chamayou, *A Theory of the Drone* (New York: New Press, 2015), 15.

68. Parikka, *Operational Images*, 31–32. The usage of *operations* here is similar to Nordin and Öberg's treatment of "war as process," the step-by-step sequences needed to conduct warfare. However as my interest is in the end-to-end operations of military targeting, I expand the notion of "process" across much larger timescales and distribute it along far greater spatiotemporal horizons—from historical emergence to sociotechnical innovation to political impact and back again focusing on epistemic operations rather than the warfare per se. See, Nordin and Öberg, "Targeting the Ontology of War."

69. See for instance, Kathrin Friedrich and A. S. Aurora Hoel, "Operational Analysis: A Method for Observing and Analyzing Digital Media Operations," *New Media & Society* 25, no. 1 (2023): 50–71; A. S. Aurora Hoel, "Styles of Objectivity in Scientific Instrumentation," in *The Oxford Handbook of Philosophy of Technology*, ed. S. Vallor (Oxford, UK: Oxford University Press, 2021); Parikka, *Operational Images*.

70. Hoel, "Styles of Objectivity."

71. Massumi, *Ontopower*, 219.

72. Friedrich and Hoel, "Operational Analysis."

73. See, for instance, Bruno Latour, *Science in Action: How to Follow Scientists and Engineers through Society* (Cambridge, MA: Harvard University Press, 1987).

74. Esmé Bosma, Marieke de Goede, and Polly Pallister-Wilkins, "Introduction: Navigating Secrecy in Security Research," in *Secrecy and Methods in Security Research: A Guide to Qualitative Fieldwork*, ed. Esmé Bosma, Marieke de Goede, and Polly Pallister-Wilkins (New York: Routledge, 2019).

75. Gilles Deleuze and Felix Guttari, *A Thousand Plateaus: Capitalism and Schizophrenia*, trans. Brian Massumi (London: Bloomsbury Academic, 2004), 12.

76. Jairus Victor Grove, *Savage Ecology: War and Geopolitics at the End of the World* (Durham, NC: Duke University Press, 2019), 16.

77. Luciana Parisi, "Speculation: A Method for the Unattainable," in *Inventive Methods: The Happening of the Social*, ed. Celia Lury and Nina Wakeford (London: Routledge, 2012).

78. The term "ecology of operations" is taken from Bernard Geoghegan and his work on how computer screens distributed information in an ecology among humans, computational instruments and the environment. Bernard Dionysius Geoghegan, "An Ecology of Operations: Vigilance, Radar, and the Birth of the Computer Screen," *Representations* 147, no. 1 (2019).

79. Hoel, "Styles of Objectivity."

80. Susan Leigh Star, "Introduction," in *Ecologies of Knowledge: Work and Politics in Science and Technology*, ed. Susan Leigh Star (New York: SUNY Press, 1995).

81. Maria Puig de la Bellacasa, "Ecological Thinking, Material Spirituality, and the Poetics of Infrastructure," in *Boundary Objects and Beyond: Working with Leigh Star*, ed. G. C. Bowker et al. (Cambridge, MA: MIT Press, 2016).

82. Massumi, *Ontopower*, 200–201, 232.

83. Hoel, "Styles of Objectivity," 111.

84. Peter Galison, "The Ontology of the Enemy: Norbert Wiener and the Cybernetic Vision," *Critical Inquiry* 21, no. 1 (1994).

Chapter 3

1. Phillip S. Meilinger, *10 Propositions Regarding Air Power* (Washington, DC: Office of Air Force History, 1995), 20.

2. Christopher Coker, "Targeting in Context," in *Targeting: The Challenges of Modern Warfare*, ed. Paul A. L. Ducheine, Michael N. Schmitt, and Frans P. B. Osinga (The Hague: T.M.C. Asser Press, 2016).

3. Caren Kaplan, *Aerial Aftermaths: Aerial Aftermaths: Wartime from Above* (Durham, NC: Duke University Press, 2018).

4. Frédéric Mégret, "War and the Vanishing Battlefield," *Loyola University Chicago International Law Review* 9, no. 1 (2012): 131–155.

5. Peter Adey, Mark Whitehead, and Alison J. Williams, "Introduction: Air-Target: Distance, Reach and the Politics of Verticality," *Theory, Culture & Society* 28, no. 7–8 (2011).

6. Hunter Heyck, *Age of System: Understanding the Development of Modern Social Science* (Baltimore, MA: Johns Hopkins University Press, 2015). While Heyck locates the "age of system" in the social sciences to the interwar and post-WWII era, the idea of system and the role of systems thinking in (re)shaping modern knowledge have a much broader and longer genealogy. On this, see for instance, Clifford Siskin, *System: The Shaping of Modern Knowledge* (Cambridge, MA: MIT Press, 2015).

7. Paul N. Edwards, *The Closed World: Computers and the Politics of Discourse in Cold War America* (Cambridge, MA: MIT Press, 1996).

8. John A. Warden, "The Enemy as a System," *Airpower Journal* 9, no. 1 (1995).

9. Derek Gregory, "Gabriel's Map: Cartography and Corpography in Modern War," in *Geographies of Knowledge and Power. Knowledge and Space*, ed. Peter Meusburger, Derek Gregory, and Laura Suarsana (Dordrecht, The Netherlands: Springer, 2015).

10. Thomas Hippler, *Governing from the Skies: A Global History of Aerial Bombing* (London: Verso, 2017), 51.

11. Maurer Maurer, ed., *The U.S. Air Service in World War I* (Washington, DC: Government Printing Office, 1978), 501–502.

12. Maurer, *The U.S. Air Service*, 502. Emphasis added.

13. Basil Henry Liddell Hart, *Paris: Or, The Future of War* (New York: Garland, 1972), 36–37.

14. John Glock, "The Evolution of Air Force Targeting," *Air Power Journal* 8, no. 3 (1994): 148.

Notes to Chapter 3

15. Hippler, *Governing from the Skies*, x–xi.

16. David E. Omissi, *Air Power and Colonial Control: The Royal Air Force, 1919–1939* (Manchester, UK: Manchester University Press, 1990).

17. For an excellent reading of the biopolitical dimensions of RAF colonial air surveys, see Peter Adey, *Aerial Life: Spaces, Mobilities, Affects* (West Sussex, UK: Wiley-Blackwell, 2010). See also, Neocleous, "Air Power as Police Power," 581.

18. Heyck, *Age of System*, 5.

19. Heyck, *Age of System*, 14.

20. Heyck, *Age of System*, 10.

21. Giulio Douhet, *The Command of the Air*, trans. Dino Ferrari (Washington, DC: Office of Air Force History and Museums Program, 1998 [1921]).

22. Haywood S. Hansell, *The Strategic Air War against Germany and Japan* (Washington, DC: Office of Air Force History, 1986), 10.

23. Hansell, *The Strategic Air War*, 11.

24. Phil M. Haun, ed., *Lectures of the Air Corps Tactical School and American Strategic Bombing in World War II* (Lexington: University Press of Kentucky, 2019), 84.

25. Tami Davis Biddle, *Rhetoric and Reality in Air Warfare: The Evolution of British and American Ideas about Strategic Bombing, 1914–1945* (Princeton, NJ: Princeton University Press, 2002), 38–39.

26. Biddle, *Rhetoric and Reality*, 139–140.

27. Hansell, *The Strategic Air War*, 13.

28. J. F. C. Fuller, *The Reformation of War* (London: Hutchinson & Company, 1923), 100.

29. Glock, "The Evolution of Air Force Targeting," 154.

30. Stephen Graham, "Demodernizing by Design: Everyday Infrastructure and Political Violence," in *Violent Geographies: Fear, Terror, and Political Violence*, ed. Derek Gregory and Allan Pred (New York: Routledge, 2007).

31. Glock, "The Evolution of Air Force Targeting," 155.

32. Frans P. B. Osinga and Mark P. Roorda, "From Douhet to Drones, Air Warfare, and the Evolution of Targeting," in *Targeting: The Challenges of Modern Warfare*, ed. Paul A. L. Ducheine, Michael N. Schmitt, and Frans P. B. Osinga (The Hague: T.M.C. Asser Press, 2016), 33.

33. Galison, "War against the Center," *Grey Room*, no. 4, Summer 2001 (2001): 8. Although *operations research* (OR) and *systems analysis* are often used interchangeably, I use them here as slightly separate forms of analysis. I see OR as the development and application of particular analytical methods to quantify military operations for the purpose of optimizing, formalizing, and modeling war to make decision-making more effective. In contrast, I see systems analysis, as developed in the interbellum years through the industrial-web theory, as more concerned with how

things hang together rather than how to operationalize military tactics. From this perspective, I would argue that Galison's "functional destruction" is more systems analysis than OR or operations analysis. While both are technical processes of abstractions that seek to model the world and to rationalize the violence that humans are subjected to, OR is more about optimizing your own warfighting, while systems analysis is more about the production of the Other.

34. William T. Walsh, "Strategic Target Analysis," *Photogrammetric Engineering* 14, no. 4 (1948): 507.

35. Walsh, "Strategic Target Analysis," 508.

36. Walt W. Rostow, "The Beginnings of Air Targeting," *Studies in Intelligence* 7, no. 1 (1963): A2.

37. Walsh, "Strategic Target Analysis," 509.

38. Brian P. Ballew, *The Enemy Objectives Unit in World War II: Selecting Targets for Aerial Bombardment that Support the Political Purpose of War* (Lucknow, India: Lucknow Books, 2014).

39. Rostow, "The Beginnings of Air Targeting,"; Brian Vlaun, *Selling Schweinfurt: Targeting, Assessment, and Marketing in the Air Campaign against Germany* (Annapolis, MD: Naval Institute Press, 2020).

40. Phillip S. Meilinger, "A History of Effects-Based Air Operations," *The Journal of Military History* 71, no. 1 (2007): 159.

41. As the US Bombing Survey states, "Conventionally the air forces designated as 'the target area' a circle having a radius of 1000 feet around the aiming point of attack. While accuracy improved during the war, Survey studies show that, in the over-all, only about 20% of the bombs aimed at precision targets fell within this target area." US Air University, *The United States Strategic Bombing Surveys: European War, Pacific War* (Maxwell Air Force Base, AL: Air University Press, 1987), 13.

42. Uta Hohn, "The Bomber's Baedeker-Target Book for Strategic Bombing in the Economic Warfare against German Towns 1943–1945," *GeoJournal* 34, no. 213–230 (1994): 213.

43. Quoted in Hohn, "The Bomber's Baedeker," 214.

44. US Air University, *The United States Strategic Bombing Surveys*, 14.

45. On moral and airpower, see Ben Anderson, "Targeting Affective Life from Above: Moral and Air Power," in *From Above: War, Violence and Verticality*, ed. Peter Adey, Mark Whitehead, and Alison Williams (Oxford, UK: Oxford University Press, 2014).

46. Coward, "On Networks and Nodes," 99.

47. US Air University, *The United States Strategic Bombing Surveys*, 36.

48. Rostow, "The Beginnings of Air Targeting."

49. Galison, "War against the Center," 8.

50. Quoted in Glock, "The Evolution of Air Force Targeting," 156.

Notes to Chapter 3

51. Walsh, "Strategic Target Analysis," 507.

52. Lynn Eden, *Whole World on Fire: Organizations, Knowledge, and Nuclear Weapons Devastation* (Ithaca, NY: Cornell University Press, 2006), 96–97.

53. Walsh, "Strategic Target Analysis."

54. Central Intelligence Agency (CIA), Air Force Intelligence, General CIA Records, 4. Unknown date. Presumably early 1950s, due to reference to atomic and not nuclear weapons and UN forces in Korea.

55. James T. Lowe quoted in Stephen Collier and Andrew Lakoff, "The Bombing Encyclopedia of the World," *Limn* 6 (2016).

56. James T. Lowe quoted in Eden, *Whole World on Fire*, 103.

57. Collier and Lakoff, "The Bombing Encyclopedia of the World."

58. CIA, Air Force Intelligence, 3.

59. CIA, Air Force Intelligence, 1.

60. CIA, Air Force Intelligence, 3. Emphasis in original.

61. CIA, Air Force Intelligence, 3. Emphasis in original.

62. Dino Brugioni, *Eyes in the Sky: Eisenhower, the CIA, and Cold War Aerial Espionage* (Annapolis, MD: Naval Institute Press, 2010), 44.

63. All quotes in this paragraph, CIA, Air Force Intelligence, 4.

64. Collier and Lakoff, "The Bombing Encyclopedia of the World."

65. Collier and Lakoff, "The Bombing Encyclopedia of the World."

66. Stephen J. Collier and Andrew Lakoff, "Vital Systems Security: Reflexive Biopolitics and the Government of Emergency," *Theory, Culture & Society* 32, no. 2 (2015).

67. Collier and Lakoff, "The Bombing Encyclopedia of the World."

68. John Durham Peters, *The Marvelous Clouds: Toward a Philosophy of Elemental Media* (Chicago: University of Chicago Press, 2015).

69. Elliott Child, "Through the Wringer: Mass Interrogation and United States Air Force Targeting Intelligence in the Early Cold War," *Political Geography* 75 (2019): 102052.

70. Child, "Through the Wringer," 10.

71. Child, "Through the Wringer," 12.

72. Galison, "War Against the Center," 231.

73. U.S. Army Special Documents Section, *Outline of Lecture to be Given to the General Council on SDS Operations* (December 16, 1946), 2.

74. Central Intelligence Agency (CIA), *Memorandum for Standing Committee Members of the IAC, Industrial Register Cards* (Langley, VA: CIA, April 15, 1949).

75. Central Intelligence Agency (CIA), *History of Office of Collection and Dissemination*, DRAFT (Langley, VA: CIA January 1, 1952), 2.

76. Brugioni, *Eyes in the Sky*, 44.

77. Central Intelligence Agency (CIA), *Memorandum for the Record, Industrial Register Survey* (Project 5–62a), Annex A, "How Material Is Indexed and Filed" (Langley, VA: CIAJanuary16, 1956), 7, 10.

78. Central Intelligence Agency, *History of Office*, 4, 7.

79. Collier and Lakoff, "The Bombing Encyclopedia of the World."

80. Outten J. Clinard, "Developments in Air Targeting: Data Handling Techniques," *Studies in Intelligence* 3, no. 2 (1959): 97.

81. Clinard, "Developments in Air Targeting," 95.

82. Clinard, "Developments in Air Targeting," 97.

83. Clinard, "Developments in Air Targeting," 97.

84. Clinard, "Developments in Air Targeting," 100.

85. Clinard, "Developments in Air Targeting," 100.

86. Clinard, "Developments in Air Targeting," 100.

87. Clinard, "Developments in Air Targeting," 101.

88. Clinard, "Developments in Air Targeting," 101.

89. Kevin McSorley, "Archives of Enmity and Martial Epistemology," in *(W)Archives: Archival Imaginaries, War, and Contemporary Art*, ed. Daniela Agostinho et al. (Berlin: Sternberg Press, 2021)

90. McSorley, "Archives of Enmity," 102.

91. McSorley, "Archives of Enmity," 102.

92. Jeremy Packer and Joshua Reeves, "Romancing the Drone: Military Desire and Anthropophobia from SAGE to Swarm," *Canadian Journal of Communication* 38, no. 3 (2013): 312.

93. Andrew Pickering, *The Mangle of Practice: Time, Agency, and Science* (Chicago: University of Chicago Press, 1995).

94. On the development of Big L systems, see Edwards, *The Closed World*, 107.

95. Kenneth T. Johnson, "Developments in Air Targeting: Progress and Future Prospects," *Studies in Intelligence* 3, no. 3 (1959): 57.

96. U.S. Congress House Committee on Appropriations, Department of Defense Appropriations for 1963, 442 (Washington, DC: U.S. Congress 1962).

97. Johnson, "Progress and Future Prospects," 58.

98. Johnson, "Progress and Future Prospects," 59.

99. Johnson, "Progress and Future Prospects," 60.

100. Johnson, "Progress and Future Prospects," 59–60.

101. Johnson, "Progress and Future Prospects," 55.

102. Heyck, *Age of System*, 167.

103. On immutable mobiles, see for instance, Bruno Latour, *Science in Action: How to Follow Scientists and Engineers through Society* (Cambridge, MA: Harvard University Press, 1987).

104. Heyck, *Age of System*, 170.

105. Susan Leigh Star and James R. Griesemer, "Institutional Ecology, 'Translations' and Boundary Objects: Amateurs and Professionals in Berkeley's Museum of Vertebrate Zoology, 1907–39," *Social Studies of Science* 19, no. 3 (1989): 387–420.

106. See, for instance, Donna Haraway, *Modest_witness@ Second_Millennium. Female_Man©_Meets_Oncomouse™: Feminism and Technoscience* (London: Routledge, 1997); John Law and Vicky Singleton, "Object Lessons," *Organization* 12, no. 3 (2005): 331–355; Marianne de Laet and Annemarie Mol, "The Zimbabwe Bush Pump: Mechanics of a Fluid Technology," *Social Studies of Science* 30, no. 2 (2000): 225–263.

107. Donald A. MacKenzie, *An Engine, Not a Camera: How Financial Models Shape Markets* (Cambridge, MA: MIT Press, 2006).

108. Heyck, *Age of System*, 169.

109. Madeline Akrich, "The De-Scription of Technical Objects." In *Shaping Technology/Building Society: Studies in Sociotechnical Change*, ed. Wiebe E. Bijker and John Law (Cambridge, MA: MIT Press, 1992), 205–224.

110. Clinard, "Developments in Air Targeting," 96.

111. Robert H. Adams, "Developments in Air Targeting: The Air Battle Model," *Studies in Intelligence* 2, no. 2 (1958): 13.

112. Johnson, "Progress and Future Prospects," 60.

113. Adams, "Developments in Air Targeting," 15.

114. Strategic Air Command, *Strategic Air Command Atomic Weapons Requirements Study for 1959 . . . SM 129–56* (Strategic Air Command, June 15, 1959), 1.

115. Adams, "Developments in Air Targeting," 15.

116. Adams, "Developments in Air Targeting," 15.

117. Adams, "Developments in Air Targeting," 19.

118. Adams, "Developments in Air Targeting," 19.

119. Johnson, "Progress and Future Prospects," 54.

120. Johnson, "Progress and Future Prospects," 55.

121. Quoted in Sharon Ghamari-Tabrizi, "Simulating the Unthinkable: Gaming Future War in the 1950s and 1960s," *Social Studies of Science* 30, no. 2 (2000): 195.

122. Johnson, "Progress and Future Prospects," 60.

123. Adams, "Developments in Air Targeting," 25.

124. Adams, "The Air Battle Model," 25–26.

125. Ghamari-Tabrizi, "Simulating the Unthinkable."

126. Johnson, "Progress and Future Prospects," 61.

127. Henry T. Nash, "The Bureaucratization of Homicide," *Bulletin of the Atomic Scientists* 36, no. 4 (1980): 22.

128. Nash, "The Bureaucratization of Homicide," 23.

129. Pickering, "Material Culture and the Dance of Agency."

130. Ghamari-Tabrizi, "Simulating the Unthinkable," 163.

131. Walsh, "Strategic Target Analysis," 512.

132. Robert W. Leavitt, "Developments in Air Targeting: The Military Resource Model," *Studies in Intelligence* 2, no. 1 (1958): 52.

133. On the rise and fall of quantitatively oriented military war-gaming and simulations, see Aggie Hirst, "Wargames Resurgent: Military Gaming and Virtual War from Recruitment to Rehabilitation," *International Studies Quarterly* 66, no. 3 (2022). For a highly interesting take on "postquantitative" challenges to OR, see Aggie Hirst, "States of Play: Evaluating the Renaissance in US Military Wargaming," *Critical Military Studies* 8, no. 1 (2020): 1–21.

134. Orit Halpern, *Beautiful Data: A History of Vision and Reason since 1945* (Durham, NC: Duke University Press, 2015); Galison, "The Ontology of the Enemy."; Edwards, *The Closed World*; Geoghegan, "An Ecology of Operations."

135. Peter Galison, "The Ontology of the Enemy: Norbert Wiener and the Cybernetic Vision," *Critical Inquiry* 21, no. 1 (1994), 230, 31.

136. For a discussion of structure versus process in cybernetics, see Halpern, *Beautiful Data*, 42–44.

137. Galison, "The Ontology of the Enemy," 256.

138. Leavitt, "Developments in Air Targeting," 52.

139. William T. Walsh, "Strategic Target Analysis," *Photogrammetric Engineering* 14, no. 4 (1948): 507.

140. Peter Galison, "War against the Center"; Stephen J. Collier and Andrew Lakoff, "Vital Systems Security: Reflexive Biopolitics and the Government of Emergency," *Theory, Culture & Society* 32, no. 2 (2015).

141. Galison, "War against the Center."

142. Collier and Lakoff, "Vital Systems Security."

143. Collier and Lakoff, "Vital Systems Security."

144. Claudia Aradau, "Security That Matters: Critical Infrastructure and Objects of Protection," *Security Dialogue* 41, no. 5 (2010); Nigel J. Thrift, *Non-Representational Theory: Space, Politics, Affect* (London: Routledge, 2008).

145. Kevin McSorley, "Archives of Enmity."

146. Paul Edwards, "Infrastructure and Modernity: Scales of Force, Time, and Social Organization in the History of Sociotechnical Systems," in *Modernity and Technology*, ed. T. J. Misa, P. Brey, and A. Feenberg (Cambridge, MA: MIT Press, 2002).

147. John Davies, Alexander Kent, and James Risen, *The Red Atlas: How the Soviet Union Secretly Mapped the World* (Chicago: University of Chicago Press, 2017).

148. See Donald A. MacKenzie, *Inventing Accuracy: A Historical Sociology of Nuclear Missile Guidance* (Cambridge, MA: MIT Press, 1993), on the mutual imbrications of technosocial imaginaries, technologies, and strategies that were neither inevitable nor "natural."

149. George H. W. Bush, "Address before a Joint Session of the Congress on the Persian Gulf Crisis and the Federal Budget Deficit," news release, September 11, 1990.

150. Thomas A. Keaney and Eliot A. Cohen, *Gulf War Air Power Survey Summary Report* (Washington, DC: 1993).

151. For example, electrical power plants with 88% of the grid shut down. Keaney and Cohen, *Gulf War Air Power Survey*, 118.

152. John A. Warden, *The Air Campaign* (Washington, DC: National Defense University Press, 1988).

153. Barry Watts, *The Maturing Revolution in Military Affairs* (Washington, DC: Center for Strategic and Budgetary Assessments, 2011).

154. Eliot A. Cohen, "A Revolution in Warfare," *Foreign Affairs* 75, no.2 (1996): 37–54.

155. Osinga and Roorda, "From Douhet to Drones," 50

156. David A. Deptula, *Effects-Based Operations: Change in the Nature of Warfare* (Arlington, Virginia: Aerospace Education Foundation, 2001).

157. Paul K. Davis, *Effect-Based Operations: A Grand Challenge for the Analytical Community*, (Santa Monica, CA: RAND, 2001), 7.

158. Osinga and Roorda, "From Douhet to Drones," 51.

159. Osinga and Roorda, "From Douhet to Drones," 50.

160. On NCW, see for instance, William A. Owens, "The Emerging Systems of Systems," *Proceedings* 121, no.5 (May 1995): 35–39; Michael C. Fowler et al., "Network Centric Warfare: Developing and Leveraging Information Superiority," *Naval War College Review* 53, no. 2 (2000); David S. Alberts and Richard E. Hayes, *Power to the Edge: Command, Control in the Information Age*, (Washington, DC: CCRP, 2003).

161. US Joint Chiefs of Staff, *Enabling the Joint Vision: Information Superiority* (Washington DC: U.S. JCS, 2000), 1.

162. Tirpak, "Find, Fix, Track, Target, Engage, Assess." *Air Force Magazine*, July 1 (2000).

163. Brian Massumi, "Perception Attack: Brief on War Time," *Theory & Event* 13, no. 3 (2010).

164. Harlan Ullman and James P. Wade, *Shock and Awe: Achieving Rapid Dominance* (Washington, DC: NDU Press, 1996), 29.

165. Ullman and Wade, *Shock and Awe*, 83.

166. Massumi, "Perception Attack."

167. Lucy Suchman, "Imaginaries of Omniscience: Automating Intelligence in the US Department of Defense," *Social Studies of Science* 53, no 5 (2022): 761–786.

168. See, for instance, John Boyd, "The Essence of Winning and Losing" (unpublished briefing, 1996). https://slightlyeastofnew.com/wp-content/uploads/2010/03/essence_of_winning_losing.pdf.

169. Antoine Bousquet, *The Eye of War: Military Perception from the Telescope to the Drone* (University of Minnesota Press, 2018), 193.

170. Mégret, "War and the Vanishing Battlefield," 8.

171. Derek Gregory, "The Everywhere War." *Geographical Journal* 177 no. 3 (2011), 238–250.

172. Dan Öberg, "Requiem for the Battlefield: The Eye of War Symposium," The Disorder of Things, January 13, 2019, https://thedisorderofthings.com/2019/01/13/requiem-for-the-battlefield/.

173. Astrid H. M. Nordin and Dan Öberg, "Targeting the Ontology of War: From Clausewitz to Baudrillard," *Millennium: Journal of International Studies* 43, no. 2 (2015): 405.

174. Nicola Perugini and Neve Gordon, "Distinction and the Ethics of Violence: On the Legal Construction of Liminal Subjects and Spaces," *Antipode* 49, no. 5 (2017), 2.

175. Graham, "Demodernizing by Design."

176. David A. Deptula et al., *Restoring America's Military Competitiveness: Mosaic Warfare* (Arlington, VA: Mitchell Institute for Aerospace Studies, September 2019); DARPA, "DARPA Tiles Together a Vision of Mosaic Warfare: Banking on Cost-Effective Complexity to Overwhelm Adversaries," DARPA, https://www.darpa.mil/work-with-us/darpa-tiles-together-a-vision-of-mosiac-warfare.

Chapter 4

1. Gordon L. Rottman, "Tactics in a Different War: Adapting US Doctrine," in *Rolling Thunder in a Gentle Land: The Vietnam War Revisited*, ed. Andrew A. Wiest (Oxford, UK: Osprey, 2006), 230–231.

2. George A. Carver, "The Faceless Viet Cong," *Foreign Affairs* 44, no. 3 (April 1966): 347–372.

3. In military doctrine, order of battle is defined as "the identification, strength, command structure, and disposition of the personnel, units, and equipment of any military force" and is often used as a template for planning operations against enemy units on the battlefield. US Joint Chiefs of Staff, "Joint Intelligence Preparation of the Operational Environment," Joint Publication 2-01.3, May 21, 2014. GL-7

4. Strategic bombing still played a prominent role in the Vietnam war, in particular, after the restrictions on fixed targets in the North were relaxed by President Nixon for the bombing campaign known as Linebacker II, in the later years of the war. See Mark Clodfelter, *The Limits of Air Power: The American Bombing of North Vietnam* (New York: Free Press, 1989).

5. Leo McCann, "'Killing Is Our Business and Business is Good': The Evolution of 'War Managerialism' from Body Counts to Counterinsurgency," *Organization* 24, no. 4 (2017): 491–515.

6. Donald Fisher Harrison, "Computers, Electronic Data, and the Vietnam War," *Archivaria* 26 (Summer 1988): 20.

7. McCann, "'Killing Is Our Business and Business Is Good.'"

8. James William Gibson, *The Perfect War: Technowar in Vietnam* (Boston: Atlantic Monthly Press, 1986).

9. Harrison, "Computers, Electronic Data, and the Vietnam War," 20.

10. Jeremy Packer, "Screens in the Sky: SAGE, Surveillance, and the Automation of Perceptual, Mnemonic, and Epistemological Labor," *Social Semiotics* 23, no. 2 (2013): 173–195.

11. Arturo Rosenblueth, Norbert Wiener, and Julian Bigelow, "Behavior, Purpose and Teleology," *Philosophy of Science* 10, no. 1 (1943): 18–24; Norbert Wiener, *Cybernetics: Or, Control and Communication in the Animal and the Machine* (Cambridge, MA: MIT Press, 1948); Steve J. Heims, *John Von Neumann and Norbert Wiener: From Mathematics to the Technologies of Life and Death* (Cambridge, MA: MIT Press, 1980).

12. Paul N. Edwards, *The Closed World: Computers and the Politics of Discourse in Cold War America* (Cambridge, MA: MIT Press, 1996).

13. Thomas L. Ahern, *CIA and Rural Pacification in South Vietnam* (Langley, VA: Center for the Study of Intelligence, Central Intelligence Agency, 2001).

14. Oliver Belcher, "Data Anxieties: Objectivity and Difference in Early Vietnam War Computing," in *Algorithmic Life: Calculative Devices in a Digital Age*, ed. Louise Amoore and Vohla Piotukh (London: Routledge, 2016), 174.

15. Oliver Belcher, "Sensing, Territory, Population: Computation, Embodied Sensors, and Hamlet Control in the Vietnam War," *Security Dialogue* 50, no. 5 (2019), 422.

16. Robert W. Komer, *Impact of Pacification on Insurgency in South Vietnam* (Santa Monica, CA: RAND, 1970), 8.

17. Komer, *Impact of Pacification*, 8. Emphasis in original

18. Belcher, "Sensing, Territory, Population," 420.

19. Erwin R. Brigham, "Pacification Measurement," *Military Review* L, no. 5 (May 1970), 51.

20. In the 1971 HES Command Manual, the HEW was replaced by the HES question-set and its accompanying HES ledger cards, but their intended use is the same. See Appendix B in Civil Operations and Rural Development Support (CORDS), Research and Analysis Directorate, "Hamlet Evaluation System (HES) Command Manual," document no. DAR R70–79 CM-01B, Military Assistance Command Vietnam, September 1, 1971.

21. Each hamlet was given a unique geographical identifier based on its Universal Transverse Mercator (UTM) grid coordinate—a standard grid system used to identify geographic points of areas. This became known as the HES ID.

22. Ahern, *CIA and Rural Pacification*, 419. According to Ahern, this HEW sample is from ca. 1968. Magnifying effect added.

23. For a full list of the indicators see Brigham, "Pacification Measurement," 49.

24. Robert W. Komer, *Organization and Management of the "New Model": Pacification Program—1966–1969* (Santa Monica, CA: RAND (1970), 199.

25. Brigham, "Pacification Measurement," 51.

26. Ithiel de Sola Pool et al., *Hamlet Evaluation System Study (HES) ACG 60F* (San Fransisco: Department of the Army, Army Concept Team in Vietnam, May 1, 1968), 53.

27. de Sola Pool et al., *Hamlet Evaluation System Study*, 2.

28. de Sola Pool et al., *Hamlet Evaluation System Study*, 74–75.

29. de Sola Pool et al., *Hamlet Evaluation System Study*, 76.

30. Komer, *Impact of Pacification*, 9.

31. Komer, *Impact of Pacification*, 9.

32. Thomas C. Thayer, *A Systems Analysis View of the Vietnam War: 1965–1972, The Situation in Southeast Asia*, Vol. 1, Assistant Secretary of Defense (Systems Analysis) (Washington, DC: Office of the Secretary of Defense 1975), 37.

33. CORDS, "HES Command Manual," 2.

34. US National Archives, "Electronic Records Relating to the Vietnam War." https://www.archives.gov/research/military/vietnam-war/electronic-data-files

35. Todd S. Bacastow et al., "From Layers to Objects: Evolving the GEOINT Analytic Tradecraft," in *The State and Future GEOINT 2017* (US Geospatial Intelligence Foundation, 2017). https://medium.com/the-state-and-future-of-geoint-2017-report/from-layers-to-objects-evolving-the-geoint-analytic-tradecraft-31698f932c61.

36. Thayer, *A Systems Analysis View of the Vietnam War*, 37.

37. Ahern, *CIA and Rural Pacification*, 406.

38. Komer, *Organization and Management*, 200.

Notes to Chapter 4

39. Komer, *Impact of Pacification*, 9–10.

40. Stew Magnuson, "Military 'Swimming in Sensors and Drowning in Data,'" National Defense, January 1, 2010. https://www.nationaldefensemagazine.org/articles/2009/12/31/2010january-military-swimming-in-sensors-and-drowning-in-data.

41. Brigham, "Pacification Measurement," 51.

42. Brigham, "Pacification Measurement," 54.

43. Brigham, "Pacification Measurement," 54, 55. Emphasis added.

44. Brigham, "Pacification Measurement," 55.

45. Belcher, "Data Anxieties," 175–176.

46. James C. Scott, *Seeing Like a State: How Certain Schemes to Improve the Human Condition Have Failed* (New Haven, CT: Yale University Press, 1998).

47. Belcher, "Sensing, Territory, Population," 419. Emphasis in original.

48. Belcher, "Sensing, Territory, Population," 419. Emphasis in original.

49. Leo Breiman, "Statistical Modeling: The Two Cultures," *Statistical Science* 16, no. 3 (2001), 202.

50. Breiman, "Statistical Modeling," 202.

51. Donna Haraway, "Situated Knowledges: The Science Question in Feminism and the Privilege of Partial Perspective," *Feminist Studies* 14, no. 3 (1988): 575–599.

52. Vivienne Jabri, *War and the Transformation of Global Politics* (London: Palgrave, 2007), 30.

53. Edwards, *The Closed World*, 12.

54. Karin Knorr Cetina, "Culture in Global Knowledge Societies: Knowledge Cultures and Epistemic Cultures," *Interdisciplinary Science Reviews* 32, no. 4 (2007): 364.

55. Dave Young, "Computing War Narratives: The Hamlet Evaluation System in Vietnam," *APRJA* 6, no. 1 (2017): 50–63.

56. U.S. Department of the Army, *Combat Intelligence*, FM 30-5 (Washington, DC: Headquarters, Department of the Army, 1973).

57. Central Intelligence Agency (CIA), "The Intelligence Attack on the Viet Cong Infrastructure," *Intelligence Memorandum* (Langley VA, CIA, May 23, 1967), 1.

58. Central Intelligence Agency (CIA), "The Intelligence Attack, 1.

59. Central Intelligence Agency (CIA), "The Intelligence Attack."

60. George A. Carver, *The Attack on the Communist (Viet Cong) Organization and Its Supporters, Particularly at the Village and Hamlet Level*, Vol. V, Foreign Relations of the United States, 1964–1968 (Washington DC: Office of the Historian 1967).

61. Carver, *The Attack on the Communist*. Emphasis added.

62. Roberto J. Gonzalez, *American Counterinsurgency: Human Science and the Human Terrain* (Chicago: Prickly Paradigm Press, 2009).

63. Carver, *The Attack on the Communist*.

64. Joseph A. McChristian, *The Role of Military Intelligence 1965–1967*, Vietnam Studies (Washington DC: Department of the Army, Army Center for Military History, 1974), 50–52.

65. Central Intelligence Agency (CIA), "Fact Sheet: Subject: Phung Hoang/Phoenix Program in Vietnam, Purpose: To Provide Information on the above Subject for the Senate Armed Services Committee" (November 25, 1969), 2.

66. Carver, *The Attack on the Communist*.

67. Dale Andrade and James Willbanks, "CORDS/Phoenix: Counterinsurgency Lessons from Vietnam for the Future," *Military Review* (March–April 2006): 77–91.

68. Central Intelligence Agency (CIA), "Phung Hoang/Phoenix Program," 2.

69. William Rosenau and Austin M. Long, *The Phoenix Program and Contemporary Counterinsurgency* (Santa Monica, CA: RAND, 2009), 11.

70. Douglas Valentine, *The Phoenix Program* (New York: Open Road Distribution, 2000).

71. Ahern, *CIA and Rural Pacification*, 287.

72. William L. Knapp, *Phoenix/Phung Hoang and the Future: A Critical Analysis of the US/GVN Program to Neutralize the Viet Cong Infrastructure*, (Carlisle, PA: Army War College, 1971), 53.

73. U.S. Military Assistance Command Vietnam (MACV), *Phung Hoang Adviser Handbook*, (Saigon, Vietnam: Headquarters, U.S. Military Assistance Command, 1970), 3.

74. MACV, *Phung Hoang Adviser Handbook*, 9. The page numbers in the quote refer to U.S. Military Assistance Command Vietnam (MACV), "Phung Hoang Standard Operating Procedures 3" (Saigon, Vietnam: Headquarters, U.S. Military Assistance Command 1970).

75. MACV, *Phung Hoang Adviser Handbook*, 10.

76. MACV, *Phung Hoang Adviser Handbook*, 1–3.

77. MACV, *Phung Hoang Adviser Handbook*, 9.

78. Brigham, "Pacification Measurement," 55.

79. Heyck, *Age of System*.

80. On the use of SNA in the US military see David Knoke, "'It Takes a Network': The Rise and Fall of Social Network Analysis in U.S. Army Counterinsurgency Doctrine," *Connections* 33, no. 1 (2013): 1–10; and Roger MacGinty, "Social Network Analysis and Counterinsurgency: A Counterproductive Strategy?" *Critical Studies on Terrorism* 3, no. 2 (2010): 209–226.

81. Marieke de Goede, "Fighting the Network: A Critique of the Network as a Security Technology," *Distinktion: Journal of Social Theory* 13, no. 3 (2012) 215–232.

82. MACV, *Phung Hoang Adviser Handbook*, 9.

Notes to Chapter 4

83. MAVC, *Phung Hoang Adviser Handbook*, Annex D.

84. Valentine, *The Phoenix Program*, 259.

85. U.S. Military Assistance Command Vietnam (MACV), Command Manual, Phung Hoang Management Information System (PHMIS), 1969–1972 (Saigon, Vietnam: Headquarters, U.S. Military Assistance Command, Vietnam, 1972), 1.

86. MAVC, Command Manual, Phung Hoang, 1.

87. MAVC, Command Manual, Phung Hoang, 2.

88. MAVC, Command Manual, Phung Hoang, 4.

89. MAVC, Command Manual, Phung Hoang.

90. MAVC, Command Manual, Phung Hoang, Appendix G 2.

91. MAVC, Command Manual, Phung Hoang, 5.

92. MAVC, Command Manual, Phung Hoang.

93. MAVC, Command Manual, Phung Hoang, 6. Emphasis added.

94. Ansorge, *Identify and Sort*, 94–95.

95. MAVC, Command Manual, Phung Hoang, 2.

96. MAVC, Command Manual, Phung Hoang, Appendix E, 4.

97. MAVC, Command Manual, Phung Hoang, Appendix E, Figure E-5, 7.

98. U.S. National Archives, *Electronic Records Relating to the Vietnam War*.

99. On relational databases, see, M. Castelle, "Relational and Non-Relational Models in the Entextualization of Bureaucracy," *Computational Culture*, no. 3 (2013): 1–58. For an excellent treatment of relational databases in the war on terror, also see Jutta Weber, "Keep Adding. On Kill Lists, Drone Warfare and the Politics of Databases," *Environment and Planning D: Society and Space* 34, no. 1 (2016): 107–125, 115.

100. Andrew R. Finlayson, "A Retrospective on Counterinsurgency Operations: The Tay Ninh Provincial Reconnaissance Unit and Its Role in the Phoenix Program, 1969–70," *Studies in Intelligence* 51, no. 2 (2007): 59–69.

101. Quoted in Rosenau and Long, *The Phoenix Program*, 11.

102. Ahern, *CIA and Rural Pacification*, 369.

103. Vic Croizat interviewed in Mai Elliot, *RAND in Southeast Asia: A History of the Vietnam War Era* (Santa Monica, CA: RAND, 2010), 302.

104. Andrade and Willbanks, "CORDS/Phoenix: Counterinsurgency Lessons from Vietnam for the Future," 20.

105. Quoted in Ahern, *CIA and Rural Pacification*, 360.

106. MAVC, Command Manual, Phung Hoang, 2.

107. Valentine, *The Phoenix Program*.

108. U.S. Department of the Army, Field Manual 30-5, 9–11. Emphasis added.

109. Ann Finkbeiner, *The Jasons: The Secret History of the Science's Postwar Elite* (New York: Viking, 2006).

110. Berkeley Scientists and Engineers for Social and Political Action, *Science against the People: The Story of Jason*, (Berkeley, CA: Berkeley SESPA, 1972).

111. Anthony J. Tambini, *Wiring Vietnam: The Electronic Wall* (Lanham, MD: Scarecrow Press, 2007), xii.

112. Derek Gregory, "Lines of Descent," openDemocracy (November 8, 2011). https://www.opendemocracy.net/en/lines-of-descent/.

113. Lucy Suchman, "Situational Awareness: Deadly Bioconvergence at the Boundaries of Bodies and Machines," *Media Tropes* 5, no. 1 (2015): 1–24.

114. Tambini, *Wiring Vietnam*, 73.

115. Philip D. Caine, "Igloo White, July 1968–December 1969 (U)" (Hickam Air Force Base, Hawaii: HQ PACAF, Directorate, Tactical Evaluation, CHECO Division, January 10, 1970), 26.

116. Caine, Igloo White, 4. Caine also remarks that there had been experiments with unmanned relay aircraft, or drones, and plans had been made to go further with this.

117. Caine, "Igloo White," 31–32.

118. Caine, "Igloo White," 4.

119. Caine, "Igloo White," 32.

120. Tambini, *Wiring Vietnam*, 73.

121. Caine, Igloo White, 8.

122. Caine, "Igloo White," 16–17.

123. Caine, "Igloo White," 17. Emphasis added.

124. U.S. Congress Senate Committee on Armed Services Electronic Battlefield Subcommittee, Investigation into Electronic Battlefield Program, Hearings Ninety-first Congress, Second Session (Washington, DC: U.S. Government Printing Office, 1970–1971), 9.

125. Gregory, "Lines of Descent," 42.

126. U.S. Congress, Investigation into Electronic Battlefield Program, 10.

127. Seymour Deitchman, "The 'Electronic Battlefield' in the Vietnam War," *Journal of Military History* 72, no. 3 (2008): 884.

128. William Westmoreland, "New Developments in Ground Warfare." Address to the Congressional Record-Senate (October 16, 1969).

129. Deitchman, "The 'Electronic Battlefield.'"

130. Michael C. Fowler et al., "Network Centric Warfare: Developing and Leveraging Information Superiority," *Naval War College Review* 53, no. 2 (2000): 229–231; David S. Alberts and Richard E. Hayes, *Power to the Edge: Command, Control in the Information Age* (Washington, DC: CCRP, 2003).

131. Gregory, "Lines of Descent."

132. Geoghegan, "An Ecology of Operations," 62–63.

133. Suchman, "Situational Awareness."

134. Harry G. Summers, "Lessons: A Soldier's View," *The Wilson Quarterly (1976–)* 7, no. 3 (1983): 125.

135. McCann, "Killing Is Our Business and Business is Good," 501.

136. Lorraine Daston and Peter Galison, "The Image of Objectivity," *Representations*, no. 40, Special Issue: Seeing Science (Autumn, 1992): 98; see also Theodore M. Porter, *Trust in Numbers: The Pursuit of Objectivity in Science and Public Life* (Princeton, NJ: Princeton University Press, 1995).

137. Geoghegan, "An Ecology of Operations."

Chapter 5

1. Stanley McChrystal, "It Takes a Network: The New Front Line of Modern Warfare," *Foreign Policy*, no 21. Online version (February 21, 2011). https://foreignpolicy.com/2011/02/21/it-takes-a-network/.

2. Stanley McChrystal, "It Takes a Network."

3. Stanley McChrystal, "It Takes a Network."

4. Flynn, Juergens, and Cantrell, "Employing ISR," *Joint Force Quarterly* 50, no. 3 (2008), 56–61.

5. McChrystal, "It Takes a Network."

6. See for example, Chamayou, *A Theory of the Drone* (New York: New Press, 2015); Grayson, *Cultural Politics of Targeted Killing: On Drones, Counter-Insurgency, and Violence* (London: Routledge, 2016).

7. Steve Niva, "Disappearing Violence: JSOC and the Pentagon's New Cartography of Networked Warfare," *Security Dialogue* 44, no. 3 (2013): 185–202. Also see Jon R. Lindsay, "Reinventing the Revolution: Technological Visions, Counterinsurgent Criticism, and the Rise of Special Operations," *Journal of Strategic Studies* 36, no. 3 (2013): 422–453.

8. Voelz, "The Individualization of American Warfare," *Parameters* 45, no. 1 (2016): 99.

9. Among studies conducted for or by the military on this topic, see, for instance, Christopher J. Lamb and Evan Munsing, *Secret Weapon: High-Value Target Teams as an Organizational Innovation*, (Institute for National Strategic Studies, National Defense University, 2011).

10. Flynn, Juergens, and Cantrell, "Employing ISR," 56.

11. US Army, Army Techniques Publication (ATP) 3-60 Targeting (May, 2015), B-1. NATO, Allied Joint Doctrine for Joint Targeting (AJP-3.9) (2016).

12. It is unknown whether this pronunciation was adopted because of ease or if it is a play on the central idea of the F3EAD cycle to constantly feed it with information.

13. Glenn J. Voelz, "The Individualization of American Warfare," 99.

14. McChrystal, "It Takes a Network"; see also Stanley McChrystal, *Team of Teams: New Rules of Engagement for a Complex World* (New York: Portfolio/Penguin, 2015), 50. The D (disseminate) in F3EAD was later added to emphasize the need to spread and share the analysis/information widely across military branches and sectors to ensure the timely expansion of operations.

15. Mitch Ferry, "F3EA—A Targeting Paradigm for Contemporary Warfare," *Australian Army Journal* 10, no. 1 (2013): 61. Emphasis added.

16. Michael Vincent Hayden, *Playing to the Edge: American Intelligence in the Age of Terror* (New York: Penguin Press, 2016), 32.

17. John Nagle, interviewed in Stephen Grey and Dan Edge, "Kill/Capture," 2011 *Frontline* (PBS). Transcript available at https://www.pbs.org/wgbh/frontline/documentary/kill-capture/transcript/.

18. U.S. Army, ATP 3-60, B-3.

19. David Kilcullen, "Countering Global Insurgency," *Small Wars Journal* version 2.2 (2004): 40.

20. Kilcullen, "Countering Global Insurgency."

21. Vivienne Jabri, *War and the Transformation of Global Politics* (London: Palgrave, 2007).

22. David H. Petraeus, "Multi-National Force-Iraq Commander's Counterinsurgency Guidance," *Military Review* 88, no. 5 (2008): 2.

23. F3EAD advanced in conjunction with a larger conceptual framework based on "attack the network" (AtN) theories reflecting the evolution and integration of network-based analysis into military thinking. See, for instance, US Joint Forces Command, *Commander's Handbook for Attack the Network* (Suffolk, VA: Joint Warfighting Center, Joint Doctrine Support Division, 2011).

24. Ferry, "F3EA," 56. Emphasis added.

25. Ferry, "F3EA," 54.

26. Donna Haraway, "Situated Knowledges: The Science Question in Feminism and the Privilege of Partial Perspective," *Feminist Studies* 14, no. 3 (1988)575–599; Lucy Suchman, *Plans and Situated Actions: The Problem of Human-Machine Communication* (New York: Cambridge University Press, 1987).

27. Flynn, Juergens, and Cantrell, "Employing ISR," 70.

28. Glenn J. Voelz, *The Rise of iWar: Identity, Information, and the Individualization of Modern Warfare* (Carlisle, PA: US Army War College Press, 2015), xiii.

Notes to Chapter 5

29. Stew Magnuson, "Military 'Swimming in Sensors and Drowning in Data,'" *National Defense*, January 1, 2010. https://www.nationaldefensemagazine.org/articles/2009/12/31/2010january-military-swimming-in-sensors-and-drowning-in-data.

30. Grey and Edge, "Kill/Capture." Emphasis added.

31. On the use of SNA in the War on Terror, see, for instance, US Joint Forces Command, *Commander's Handbook for Attack the Network*.

32. Marieke de Goede, "Fighting the Network: A Critique of the Network as a Security Technology," *Distinktion: Journal of Social Theory* 13, no. 3 (2012), 215.

33. de Goede, "Fighting the Network," 221.

34. de Goede, "Fighting the Network," 222.

35. Robert Clark, *Intelligence Analysis: A Target-Centric Approach* (Thousand Oaks, CA: SAGE, 2017), 185.

36. de Goede, "Fighting the Network," 228.

37. Nada Bakos, *The Targeter: My Life in the CIA, Hunting Terrorists and Challenging the White House* (New York: Little, Brown and Company, 2019), 157.

38. de Goede, "Fighting the Network," 228.

39. Bakos, *The Targeter*, 157.

40. Clark, *Intelligence Analysis*, 192.

41. Clark, *Intelligence Analysis*, 191.

42. Bruno Latour, "Networks, Societies, Spheres: Reflections of an Actor-Network Theorist," *International Journal of Communication* 5 (2011): 800.

43. Eyal Weizman, "Walking through Walls: Soldiers as Architects in the Israeli-Palestinian Conflict," *Radical Philosophy* 136, March/April (2006): 8–22.

44. de Goede, "Fighting the Network," 221.

45. HVI "is a person of interest who is identified, surveilled, tracked, influenced, or engaged" (U.S. Army, ATP 3-60, B-3).

46. Greg Miller, "Plan for Hunting Terrorists Signals U.S. Intends to Keep Adding Names to Kill Lists," *Washington Post*, October 23 2012.

47. US Joint Chiefs of Staff, Special Operations, Joint Publication 3-05, A-4 (Washington, DC: US JCS 2014).

48. Flynn, Juergens, and Cantrell, "Employing ISR," 55.

49. U.S. Army, ATP 3-60, D-10.

50. U.S. Army, ATP 3-60, D-10.

51. Josh Begley, "The Drone Papers: A Visual Glossary," *The Intercept*, October 15, 2015. https://theintercept.com/drone-papers/a-visual-glossary/.

52. US Army, ATP 3-60, B-3.

53. US Army, ATP 3-60, B-1.

54. US Army, ATP 3-60, B-3.

55. US Army, ATP 3-60, B-3.

56. US Army, ATP 3-60, B-1.

57. See, for instance, National Security Agency (NSA), "Special Training on FISA, Module 3: (U) Establishing Reasonable Articulable Suspicion (RAS)," (2007). https://www.dni.gov/files/documents/1118/CLEANEDOVSC1205_L3_storyboard_v22_Final.pdf.

58. Government Communications Head Quarters (GCHQ), *HIMR Data Mining Research Problem Book* (GCHQ 2011), 12. https://www.maths.ed.ac.uk/~tl/docs/Problem-Book-Redacted.pdf.

59. National Security Agency (NSA), "Business Records (BR), FISA Course." https://snowden.glendon.yorku.ca/items/show/716.

60. See, for instance, special issue on visual analytics in; National Security Agency (NSA), "An Information Visualization Primer and Field Trip," *The Next Wave: The National Security Agency's Review of Emerging Technology* 17, no. 2 (2008).

61. US Army, ATP 3-60, B-1.

62. Grégoire Chamayou, "Oceanic Enemy: A Brief Philosophical History of the NSA," *Radical Philosophy* 191, May/June (2015): 6.

63. US Army, ATP 3-60, B-3.

64. Flynn, Juergens, and Cantrell, "Employing ISR," 58.

65. Flynn, Juergens, and Cantrell, "Employing ISR," 58.

66. G. A. Gross et al., "Application of Multi-Level Fusion for Pattern of Life Analysis." Paper presented at the 18th International Conference on Information Fusion, 2015.

67. Flynn, Juergens, and Cantrell, "Employing ISR."

68. Gross et al., "Application of Multi-Level Fusion."

69. Joseph Pugliese, *State Violence and the Execution of Law* (Abingdon, UK: Routledge, 2013), 193–194.

70. Excerpt from *First Platoon*, drawn from Annie Jacobsen, "Palantir's God's-Eye View of Afghanistan," *Wired Magazine*, January 20, 2021. Emphasis added.

71. "SIGINT is the coin of the realm" is a military saying in Afghanistan. *SIDtoday* Editor, "What's Really Going on in Afghanistan? An Interview with Brian Goodman, Recent NCR Afghanistan," *SIDtoday*, May 12, 2009.

72. Glenn Greenwald, *No Place to Hide: Edward Snowden, the NSA, and the U.S. Surveillance State* (New York: Metropolitan Books, 2014), 97. Ellen Nakashima and Joby Warrick, "For NSA Chief, Terrorist Threat Drives Passion to 'Collect it All,'" *Washington Post*, July 14, 2013.

Notes to Chapter 5

73. See, for instance, US Joint Forces Command, *Commander's Handbook for Persistent Surveillance* (Suffolk, VA: Joint Warfighting Center, Joint Doctrine Support Division, 2011).

74. National Security Agency (NSA), *(U) SIGINT Strategy, 2012–2016*, (NSA,2012): 2. https://archive.org/details/NSA-SIGINT-Strategy/page/n3/mode/2up?q=ingest

75. US Joint Chiefs of Staff, "Joint Publication 3-25," *Countering Threat Networks* (Arlington, VA: JCS, 2016), III-9.

76. Nakashima and Warrick, "For NSA Chief."

77. Jutta Weber, "Keep Adding. On Kill Lists, Drone Warfare and the Politics of Databases," *Environment and Planning D: Society and Space* 34, no. 1 (2016): 115

78. Weber, "Keep Adding," 107–125, 113.

79. Graham Harwood, "Endless War: On the Database Structure of Armed Conflict," *Rhizome*, May 17, 2014. https://rhizome.org/editorial/2014/mar/17/endless-war -database-structure-armed-conflict/.

80. Chamayou, "Oceanic Enemy," 7. Emphases in original.

81. Derek Gregory, "Dis/Ordering the Orient: Scopic Regimes and Modern War," in *Orientalism and War*, ed. Tarak Barkawi and K. Stanski (London: Hurst, 2012)

82. See, for instance, Steve Coll, "The Unblinking Stare: The Drone War in Pakistan," *New Yorker*, November 24, 2014.

83. Jim Thomas and Christopher Dougherty, *Beyond the Ramparts: The Future of U.S. Special Operations Forces* (Washington, DC: Center for Strategic and Budgetary Assessments [CSBA], 2013), 18.

84. Weber, "Keep Adding."

85. Voelz, *The Rise of iWar*, 48.

86. Hito Steyerl. "A Sea of Data: Pattern Recognition and Corporate Animism (Forked Version)." In *Pattern Discrimination*, ed. Clemens Apprich et al. (Lüneburg, Germany: Meson Press, 2018), 1–22.

87. National Security Agency (NSA), *SKYNET: Courier Detection via Machine Learning* (2012). https://snowden.glendon.yorku.ca/items/show/597.

88. National Security Agency (NSA), *SKYNET*, 2. Emphasis added.

89. See, for instance, GCHQ, *HIMR Data Mining Research*, 3.

90. Christian Grothoff and J. M. Porup, "The NSA's SKYNET Program May Be Killing Thousands of Innocent People," *arsTechnica*, 2016. https://arstechnica.com /information-technology/2016/02/the-nsas-skynet-program-may-be-killing -thousands-of-innocent-people/3/.

91. Paul Kockelman, "The Anthropology of an Equation: Sieves, Spam Filters, Agentive Algorithms, and Ontologies of Transformation," *HAU: Journal of Ethnographic Theory* 3, no. 3 (2013): 36.

92. Matteo Pasquinelli, "How a Machine Learns and Fails: A Grammar of Error for Artificial Intelligence," *Spheres*, no. 5 (2019). https://spheres-journal.org/contribution/how-a-machine-learns-and-fails-a-grammar-of-error-for-artificial-intelligence/.

93. NSA, *SKYNET*, 20.

94. John Cheney-Lippold, "A New Algorithmic Identity: Soft Biopolitics and the Modulation of Control," *Theory, Culture & Society* 28, no. 6 (2011): 164–181.

95. Gilles Deleuze, "Postscript on the Societies of Control," *October* 59, Winter (1992): 3–7.

96. Claudia Aradau and Tobias Blanke, "Governing Others: Anomaly and the Algorithmic Subject of Security," *European Journal of International Security* 3, no. 1 (2018): 1–21.

97. Grothoff and Porup, "The NSA's SKYNET Program."

98. On the dilemma between accuracy and simplicity in modeling, see Leo Breiman, "Statistical Modeling: The Two Cultures," *Statistical Science* 16, no. 3 (2001): 199–231.

99. General Mark Alexander Milley, chairman of the Joint Chiefs of Staff, quoted in the SIDtoday Editor, "What's Really Going on in Afghanistan?"

100. US Army, ATP 3-60, B-5.

101. US Army, ATP 3-60, B-5.

102. Charles Faint and Michael Harris, "F3EAD: OPS/Intel Fusion 'Feeds' the SOF Targeting Process," *Small Wars Journal* 31, no. 7 (2012). https://smallwarsjournal.com/jrnl/art/f3ead-opsintel-fusion-%E2%80%9Cfeeds%E2%80%9D-the-sof-targeting-process.

103. ISR Task Force, *ISR Support to Small Footprint CT Operations—Somalia/Yemen* (ISR Task Force Requirements and Analysis Division, 2013), 8. https://goodtimesweb.org/overseas-war/2015/small-footprint-operations-february-2013.pdf.

104. ISR Task Force, *ISR Support*, 9.

105. Thomas and Dougherty, *Beyond the Ramparts*, 18.

106. Grey and Edge, "Kill/Capture."

107. See, for instance, Derek Gregory, "From a View to a Kill: Drones and Late Modern War," *Theory, Culture & Society* 28, no. 7–8 (2011): 188–215.

108. Lamb and Munsing, *Secret Weapon*, 34.

109. Lamb and Munsing, *Secret Weapon*, 12–13.

110. Bakos, *The Targeter*, 105–106.

111. Bakos, *The Targeter*, 167.

112. Bakos, *The Targeter*, 170.

113. Frans Osinga, *Science, Strategy and War: The Strategic Theory of John Boyd* (London: Routledge, 2007).

Notes to Chapter 5

114. Bakos, *The Targeter*, 170.

115. Bakos, *The Targeter*, 171.

116. Bakos, *The Targeter*, 155.

117. Lindsay, "Target Practice: Counterterrorism and the Amplification of Data Friction." *Science, Technology, & Human Values* 42, no. 6 (2017): 1061–1099.

118. N. Katherine Hayles, "Cognitive Assemblages: Technical Agency and Human Interactions," *Critical Inquiry* 43, no. 1 (2016): 32–55, 35.

119. Ferry, "F3EA," 58.

120. Ben Anderson, "Facing the Future Enemy: US Counterinsurgency Doctrine and the Pre-insurgent," *Theory, Culture & Society* 28, no. 7–8 (2011): 216–240.

121. Brian Massumi, "Potential Politics and the Primacy of Preemption," *Theory & Event* 10, no. 2 (2007). https://muse.jhu.edu/article/218091

122. Eyal Weizman, *Hollow Land: Israel's Architecture of Occupation* (London: Verso, 2012), 308, end note 11.

123. Louise Amoore and Marieke de Goede, eds., *Risk and the War on Terror* (London: Routledge, 2008); Claudia Aradau and Rens Van Munster, "Governing Terrorism Through Risk: Taking Precautions, (un)Knowing the Future," *European Journal of International Relations* 13, no. 1 (2007): 89–115.

124. Massumi, "Potential Politics."

125. Now a key part of Australian Doctrine: Australian Army Headquarters, *Adaptive Campaigning–Future Land Operating Concept* (Canberra: Directorate of Army Research and Analysis, 2009).

126. Anne-Marie Grisogono and Alex Ryan, "Adapting C2 to the 21st Century: Operationalising Adaptive Campaigning." Paper presented at the 2007 CCRTS, Naval War College, Newport, RI, 2007 (all quotes from p. 2).

127. Grisogono and Ryan, "Adapting C2 to the 21st Century," 3.

128. Grisogono and Ryan, "Adapting C2 to the 21st Century," 3.

129. Grisogono and Ryan, "Adapting C2 to the 21st Century," 2.

130. Caroline Holmqvist, *Policing Wars: On Military Intervention in the Twenty-First Century* (Basingstoke, UK: Palgrave Macmillan UK, 2014) 37.

131. See, for instance, Mark Neocleous, *War Power, Police Power* (Edinburgh: Edinburgh University Press, 2014).

132. See Astrid H. M. Nordin and Dan Öberg, "Targeting the Ontology of War: From Clausewitz to Baudrillard," *Millennium: Journal of International Studies* 43, no. 2 (2015): 392–410.

133. Jan Bachmann, Caroline Holmqvist, and Coleen Bell, eds., *War, Police and Assemblages of Intervention* (London: Routledge, 2015).

134. Michael Dillon and Julian Reid, *The Liberal Way of War: Killing to Make Life Live* (London: Routledge, 2009); Jabri, *War and the Transformation of Global Politics.*

135. Grégoire Chamayou, *Manhunts: A Philosophical History* (Princeton, NJ: Princeton University Press, 2012).

136. Katharine Hall Kindervater, "The Emergence of Lethal Surveillance: Watching and Killing in the History of Drone Technology," *Security Dialogue* 47, no. 3 (2016), 233.

137. Paul N. Edwards, *The Closed World: Computers and the Politics of Discourse in Cold War America* (Cambridge, MA: MIT Press, 1996); Karin Knorr Cetina, "Culture in Global Knowledge Societies: Knowledge Cultures and Epistemic Cultures," *Interdisciplinary Science Reviews* 32, no. 4 (2007): 364.

138. Lindsay, "Target Practice," 1081.

139. Perhaps best summed up with what in the martial community is called "Clapper's law" (after the former director of NGA and US Director of National Intelligence James Clapper): "Everything and everybody has to be somewhere." Patrick Biltgen and Stephen Ryan, *Activity-Based Intelligence: Principles and Applications* (Boston: Artech House, 2016), 10.

140. Chamayou, *A Theory of the Drone*, 57.

141. Chamayou, *A Theory of the Drone*, 57.

142. Gregory, "The Everywhere War," *The Geographical Journal* 177, no. 3 (2011), 238–250; Lisa Parks and Caren Kaplan, eds., *Life in the Age of Drone Warfare* (Durham, NC: Duke University Press, 2017).

143. Frédéric Mégret, "War and the Vanishing Battlefield," *Loyola University Chicago International Law Review* 9, no. 1 (2012), 15.

144. For an excellent treatment of the reappearance of the battlefield through military space-time modeling, see Dan Öberg, "Requiem for the Battlefield: The Eye of War Symposium," The Disorder of Things, January 13, 2019, https://thedisorderofthings .com/2019/01/13/requiem-for-the-battlefield/

145. Hayden, *Playing to the Edge*, 333.

Chapter 6

1. This statement is known in the US intelligence community as Clapper's law. Patrick Biltgen and Stephen Ryan, *Activity-Based Intelligence: Principles and Applications* (Boston: Artech House, 2016), 10.

2. Cryptologic Support Group (CSG) was a miniature version of the NSA that could be sent to the frontlines. Members contained a mix of skill sets, from geospatial analysts to digital network intelligence (DNI) analysts in addition to SIGINT.

3. SHARKFINN is a high-capacity and high-speed "vacuum cleaner" that sweeps up all-source communication intelligence (COMINT).

4. IBM visualization software based on ELP (entity-link-property) models that would find "hidden" connections in data, creating relations and patterns through ingesting, fusing, and analyzing.

5. A program similar to Google Maps that can portray so-called real-time intelligence on digital maps and link up with targeting databases.

6. NSA Deployer. "Hollywood Special Effects? No, It's Modern-Day SIGINT," SIDtoday, July 8, 2009, https://theintercept.com/snowden-sidtoday/5987519-hollywood-special-effects-no-it-s-modern-day/.

7. National Security Agency (NSA), "Joint Document Highlights NGA and NSA Collaboration," NSA News release, December 27, 2004. https://www.nsa.gov/Press-Room/Press-Releases-Statements/Press-Release-View/Article/1635467/.

8. National Geospatial Intelligence Agency (NGA), *Geospatial Intelligence (GEOINT) Basic Doctrine* (Springfield, VA: NGA, 2018), 12.

9. Matt Alderton, "The Defining Decade of GEOINT," *Trajectory Magazine*, March 13, 2014. https://www.mattalderton.com/wp-content/uploads/2020/11/USGIF_Decade.pdf

10. James E. Heath, "RT10: Getting Information to the Front Lines in Time to Make a Difference," SIDtoday, April 18, 2007), https://theintercept.com/snowden-sidtoday/5987511-rt10-getting-information-to-the-front-lines-in/.

11. STRAP is a codeword used by British intelligence as an additional classification for particular sensitive intelligence to further restrict access to that information or document.

12. Government Communications Head Quarters (GCHQ), *GCHQ Summary of RTRG, Top Secret Strap* 1 (GCHQ, 2011), 2. Emphasis added. https://theintercept.com/document/gchq-summary-of-rtrg/.

13. Heath, *RT10.*

14. Heath, *RT10.*

15. National Security Agency (NSA), *Presentation on RTRG Analytics for Forward Users* (NSA, 2012): 16. https://theintercept.com/document/nsa-presentation-on-rtrg-analytics-for-forward-users/

16. National Security Agency (NSA), *Presentation on RTRG Analytics*, 16.

17. National Security Agency (NSA), *Presentation on RTRG Analytics*, 17.

18. National Security Agency (NSA), *Presentation on RTRG Analytics*, 16.

19. National Security Agency (NSA), *Presentation on RTRG Analytics*, 4

20. SIDtoday Editor, "What Are the Latest SIGINT Developments in Iraq and Afghanistan? An Interview with Colonel Parker Schenecker," SIDtoday, October 21, 2009. https://theintercept.com/snowden-sidtoday/5987522-what-are-the-latest-sigint-developments-in-iraq/.

21. SIDtoday is the internal newsletter for one of NSA's most important divisions, the Signals Intelligence Directorate.

22. SIDtoday Editor, "What Are the Latest."

23. All quotes from Susan Weinberger are from "The Graveyard of Empires and Big Data: The Pentagon's Secret Plan to Crowdsource Intelligence from Afghan Civilians Turned out to Be brilliant—Too brilliant," *Foreign Policy*, March 2017. https://foreignpolicy.com/2017/03/15/the-graveyard-of-empires-and-big-data/.

24. Chandler P. Atwood, "Activity-Based Intelligence: Revolutionizing Military Intelligence Analysis," *Joint Forces Quarterly* 77, no. 2nd quarter (2015): 24–29; James L. Lawrence, "Activity-Based Intelligence Coping with the 'Unknown Unknowns' in Complex and Chaotic Environments," *American Intelligence Journal* 33, no. 1 (2016): 17–25.

25. Derek Gregory, "From a View to a Kill: Drones and Late Modern War," *Theory, Culture & Society* 28, no. 7–8 (2011), 195.

26. Long, "Activity Based Intelligence."

27. For a practitioner's introduction to activity-based intelligence (ABI), see Biltgen and Ryan, *Activity-Based Intelligence*.

28. Quoted in Biltgen and Ryan, *Activity-Based Intelligence*, 25. Also see National System for Geospatial Intelligence, *Geospatial Intelligence (GEOINT) Basic Doctrine*, 34 (Springfield, VA: NGA, 2018).

29. Gabriel Miller, "Activity-Based Intelligence Uses Metadata to Map Adversary Networks," *Defense News*, July 8, 2013. https://www.ocnus.net/article.php?Activity-Based-Intelligence-Uses-Metadata-to-Map-Adversary-Networks-28011.

30. William Raetz, "A New Approach to Graph Analysis for Activity Based Intelligence." In *2012 IEEE Applied Imagery Pattern Recognition Workshop (AIPR)* (Washington, DC: IEEE, 2012): 1.

31. Biltgen and Ryan, *Activity-Based Intelligence*.

32. Kristin Quinn, "A Better Toolbox," *Trajectory Magazine*, Winter (2012): 13.

33. Atwood, "Activity-Based Intelligence," 27.

34. Quinn, "A Better Toolbox."

35. Ben Conklin, "Activity Based Intelligence: A Perilous Journey to Intelligence Integration," *Esri*, 9. n.d. https://www.esri.com/content/dam/esrisites/sitecore-archive/Files/Pdfs/library/presentations/ABI-Perilous_Journey.pdf?srsltid=AfmBOoqPHZ4xgktigq5jm_rS2MubbkO_4am2oKSao_pamosSSEPqwiL1, 9.

36. G. A. Gross et al., "Application of Multi-Level Fusion for Pattern of Life Analysis." Paper presented at the 18th International Conference on Information Fusion, 2015.

37. For an excellent analysis on the different methods used, see Claudia Aradau and Tobias Blanke, "Governing Others: Anomaly and the Algorithmic Subject of Security," *European Journal of International Security* 3, no. 1 (2018): 1–21.

38. Guansong Pang et al., "Deep Learning for Anomaly Detection: A Review," *ACM Computing Surveys* 54, no. 2 (2021): 1.

Notes to Chapter 6   257

39. Gross et al., "Application of Multi-Level Fusion."

40. Pang et al., "Deep Learning for Anomaly Detection."

41. Todd G. Myers, "GEOINT Big Data: Implementing the Right Big Data Architecture," NGA 2013 Joint GMU-AFCEA Symposium, 2013, 10.

42. Raetz, "A New Approach to Graph Analysis," 2.

43. Ian Hacking, "How Should We Do the History of Statistics?" in *The Foucault Effect: Studies in Governmentality*, ed. Graham Burchill, Colin Gordon, and Peter Miller (Chicago: University of Chicago Press, 1991), 188.

44. Grégoire Chamayou, "Patterns of Life: A Very Short History of Schematic Bodies," *Funambulist Magazine* 57 (2014). Emphasis in original. https://thefunambulist.net/editorials/the-funambulist-papers-57-schematic-bodies-notes-on-a-patterns-genealogy-by-gregoire-chamayou.

45. Michel Foucault, *Security, Territory, Population: Lectures at the Collège de France, 1977–78*, trans. Graham Burchell (London: Palgrave Macmillan, 2007), 57.

46. Michel Foucault, *Security, Territory, Population*, 63.

47. Louise Amoore and Volha Piotukh, "Life Beyond Big Data: Governing with Little Analytics," *Economy and Society* 44, no. 3 (2015): 359.

48. Aradau and Blanke, "Governing Others," 11.

49. Aradau and Blanke, "Governing Others," 3–4.

50. Chamayou, "Patterns of Life."

51. Chamayou, "Patterns of Life."

52. Hacking, "Making up People," in *Reconstructing Individualism. Autonomy, Individuality, and the Self in Western Thought*, ed. Thomas C. Heller and Christine Brooke-Rose (Stanford, CA: Stanford University Press, 1986)

53. Chamayou, "Patterns of Life."

54. Henri Lefebvre, *Rhythmanalysis: Space, Time, and Everyday Life*, trans. Stuart Elden and Gerald Moore (London: Continuum, 2014), 6.

55. Matteo Pasquinelli, "Anomaly Detection: The Mathematization of the Abnormal in the Metadata Society," panel presentation at *Transmediale Festival*, Berlin, 2015.

56. Aradau and Blanke, "Governing Others," 11.

57. Chamayou, "Patterns of Life."

58. Olga Goriunova, "The Digital Subject: People as Data as Persons," *Theory, Culture & Society* 36, no. 6 (2019): 125–145.

59. John Cheney-Lippold, *We Are Data: Algorithms and The Making of Our Digital Selves* (New York: NYU Press, 2017).

60. Aradau and Blanke, "Governing Others," 5.

61. Aradau and Blanke, "Governing Others," 9–10.

62. Pang et al., "Deep Learning for Anomaly Detection."

63. Todd S. Bacastow et al., "From Layers to Objects: Evolving the GEOINT Analytic Tradecraft," in *The State and Future GEOINT 2017* (US Geospatial Intelligence Foundation, 2017). https://medium.com/the-state-and-future-of-geoint-2017-report/from-layers-to-objects-evolving-the-geoint-analytic-tradecraft-31698f932c61.

64. The notion of technogenesis is according to Hayles: "the idea that humans and technics have coevolved together." N. Katherine Hayles, *My Mother Was a Computer: Digital Subjects and Literary Texts* (Chicago: University of Chicago Press, 2010), 10.

65. Grégoire Chamayou, "Oceanic Enemy: A Brief Philosophical History of the NSA," *Radical Philosophy* 191 (May/June, 2015): 6.

66. Luciana Parisi and Steve Goodman, "Mnemonic Control," in *Beyond Biopolitics: Essays on the Governance of Life and Death*, ed. Patricia Ticineto Clough and Craig Willse (Durham, NC: Duke University Press, 2011), 170.

67. Samuel Kinsley, "Memory Programmes: The Industrial Retention of Collective Life," *Cultural Geographies* 22, no. 1 (2015): 156.

68. Samuel Kinsley, "Memory Programmes," 165.

69. Samuel Kinsley, "Memory Programmes," 165.

70. Wolfgang Ernst, "The Archive as Metaphor: From Archival Space to Archival Time," *Open* 7, (2004): 46–53.

71. Parisi and Goodman, "6. Mnemonic Control," 171.

72. Chamayou, "Oceanic Enemy," 8.

73. Shintaro Miyazaki, "Algorhythmics: Understanding Micro-Temporality in Computational Cultures," *Computational Culture* 2, September 28 (2012), 5.

74. Claudio Coletta and Rob Kitchin, "Algorhythmic Governance: Regulating the 'Heartbeat' of a City Using the Internet of Things," *Big Data & Society* 4, no. 2 (2017): 2.

75. See, for instance, Antoinette Rouvroy and Bernard Stiegler, "The Digital Regime of Truth: From the Algorithmic Governmentality to a New Rule of Law," *La Deleuziana* 3 (2016): 6–23; John Cheney-Lippold, "A New Algorithmic Identity: Soft Biopolitics and the Modulation of Control," *Theory, Culture & Society* 28, no. 6 (2011): 164–181; Jeremy W. Crampton and Andrea Miller, "Introduction to Intervention Symposium: Algorithmic Governance," *Antipode,* May 19 (2017). https://antipodeonline.org/2017/05/19/algorithmic-governance/.

76. Coletta and Kitchin, "Algorhythmic Governance," 4.

77. Coletta and Kitchin, "Algorhythmic Governance," 5.

78. Coletta and Kitchin, "Algorhythmic Governance," 3.

79. Rob Kitchin, "The Timescape of Smart Cities," *Annals of the American Association of Geographers* 109, no. 3 (2019): 776.

80. Kitchin, "The Timescape of Smart Cities," 783.

81. Dan Öberg, "Requiem for the Battlefield: The Eye of War Symposium," The Disorder of Things, January 13, 2019, https://thedisorderofthings.com/2019/01/13/requiem-for-the-battlefield/.

82. Patrick Tucker, "The Future the US Military Is Constructing: A Giant, Armed Nervous System," *Defense One*, September 26, 2017. https://www.defenseone.com/technology/2017/09/future-us-military-constructing-giant-armed-nervous-system/141303/.

83. Benjamin H. Bratton, *The Stack: On Software and Sovereignty* (Cambridge, MA: MIT Press, 2016).

84. National Geospatial-Intelligence Agency, *Future State Vision 2018* (Springfield, VA: National Geospatial-Intelligence Agency, 2013), 2.

85. National Geospatial-Intelligence Agency, *Future State Vision 2018*, 2.

86. Edwards Abrahams et al., "Everything, Everywhere, All the Time: Now What? What Will Result from the Persistence and Depth of Geospatial Data?" *Trajectory Magazine*, February 1, 2018. https://medium.com/the-2018-state-and-future-of-geoint-report/everything-everywhere-all-the-time-now-what-fe7d4e52113f.

87. Ashley M. Richter, Rupal Mehta, and Michael Hess, "The Frontier of Multimodal Mapping: The Future of Secure, Integrated Data Visualization," *Trajectory Magazine*, January 25, 2019. https://medium.com/the-2019-state-and-future-of-geoint-report/the-frontier-of-multimodal-mapping-8247c27038cb.

88. Richter, Mehta, and Hess, "The Frontier of Multimodal Mapping."

89. Richter, Mehta, and Hess, "The Frontier of Multimodal Mapping."

90. Arturo Rosenblueth and Norbert Wiener, "The Role of Models in Science," *Philosophy of Science* 12, no. 4 (1945): 316.

91. Jacquelyn Karpovich, "NGA Director Provides Vision for Next Phase of Intel at GEOINT Symposium," *Pathfinder* 12, no. 2 (2014): 6.

92. Norbert Wiener, *The Human Use of Human Beings: Cybernetics and Society* (London: Free Association Books, 1989 [1950]), 97.

93. Karpovich, "NGA Director Provides Vision."

94. Jeannette M. Wing, "Computational Thinking," *Communications of the ACM* 49, no. 3 (2006): 33–35.

95. Nancy Rapavi, "The Right Questions: What Do We Really Need to Know?" *Pathfinder* 15, no. 3 (2017): 14–17.

96. On Whitehead, see for instance, Isabelle Stengers, *Thinking with Whitehead: A Free and Wild Creation of Concepts* (Cambridge, MA: Harvard University Press, 2011).

97. Alfred North Whitehead, *Process and Reality: An Essay in Cosmology*, ed. David Ray Griffin and Donald W. Sherburne (New York: Free Press, 1978 [1929]), 21–23.

98. Donna Haraway, "When Species Meet: Staying with the Trouble," *Environment and Planning D: Society and Space* 28, no. 1 (2010): 53–55.

99. Karen Barad, "Nature's Queer Performativity," *Qui Parle* 19, no. 2 (2011), 46

100. Lucy Suchman, "Situational Awareness: Deadly Bioconvergence at the Boundaries of Bodies and Machines," *Media Tropes* 5, no. 1 (2015): 6.

101. Luciana Parisi, "Critical Computation: Digital Automata and General Artificial Thinking," *Theory, Culture & Society* 36, no. 2 (2019), 89.

102. Massumi, *Ontopower* (Durham, NC: Duke University Press, 2015). Also see David Chandler, *Ontopolitics in the Anthropocene: An Introduction to Mapping, Sensing and Hacking* (London: Routledge, 2018).

103. Chandler, *Ontopolitics in the Anthropocene.*

104. For an overview see Dan Öberg, "Warfare as Design: Transgressive Creativity and Reductive Operational Planning," *Security Dialogue* 49, no. 6 (2018): 493–509. While Öberg focuses on the link between discourses of military design and creativity in warfare, this book is more interested in the link between military design thinking, computational logics and epistemologies, and experimentation in the production and operationalization of the enemy.

105. Louise Amoore, *The Politics of Possibility: Risk and Security Beyond Probability* (Durham, NC: Duke University Press, 2013).

106. See, for instance, Crampton and Miller, "Algorithmic Governance."

107. C. Aradau, and T. Blanke, *Algorithmic Reason: The New Government of Self and Other* (Oxford, UK: Oxford University Press, 2022).

108. Jutta Weber, "Keep Adding. On Kill Lists, Drone Warfare and the Politics of Databases," *Environment and Planning D: Society and Space* 34, no. 1 (2016): 107–125.

109. Louise Amoore and Rita Raley, "Securing with Algorithms: Knowledge, Decision, Sovereignty," *Security Dialogue* 48, no. 1 (2017): 5.

110. James C. Scott, *Seeing Like a State: How Certain Schemes to Improve the Human Condition Have Failed* (New Haven, CT: Yale University Press, 1998), 11.

111. Leo Breiman, "Statistical Modeling: The Two Cultures," ," *Statistical Science* 16, no. 3 (2001): 202.

112. Breiman, "Statistical Modeling," 205. Emphasis in original.

113. Breiman, "Statistical Modeling," 209.

114. Pang et al., "Deep Learning for Anomaly Detection," 4–5. Emphasis added.

115. Rob Kitchin, "Big Data, New Epistemologies and Paradigm Shifts," *Big Data & Society* 1, no. 1 (2014): 1–12.

116. Weber, "Keep Adding," 109. Emphasis in original.

117. Weber, "Keep Adding," 116.

Notes to Chapter 6

118. Parisi, "Critical Computation," 99.

119. Wendy Nelson Espeland and Mitchell L. Stevens, "Commensuration as a Social Process," *Annual Review of Sociology* 24 (1998): 1–12.

120. N. Katherine Hayles, *Unthought: The Power of the Cognitive Nonconscious* (Chicago: University of Chicago Press, 2017).

121. Louise Amoore, "Doubt and the Algorithm: On the Partial Accounts of Machine Learning," *Theory, Culture & Society* 36, no. 6 (2019): 3.

122. Philippe Beaulieu-Brossard and Philippe Dufort, "Introduction: Revolution in Military Epistemology: Reflexive Military Practitioners: Design Thinking and Beyond," *Journal of Military and Strategic Studies* 17, no. 4 (2017): 1–20.

123. See Weizman, "Walking Through Walls." Developed by Brigadier General (retired) Dr. Shimon Naveh after his completion of a Ph.D. at King's College London on the evolution of operational art. Naveh's thinking is heavily influenced by Soviet Operational Art from the 1920s, architecture theory, and poststructuralism, especially Deleuze and Guttari. Ofra Graicer calls the Israeli Systemic Operational Design (SOD) "the postmodern incarnation of Soviet Operational Art in western militaries." Ofra Graicer, "Self Disruption: Seizing the High Ground of Systemic Operational Design (SOD)," *Journal of Military and Strategic Studies* 17, no. 4 (2017): 21–37.

124. Beaulieu-Brossard and Dufort, "Introduction: Revolution in Military Epistemology," footnote number 3.

125. Beaulieu-Brossard and Dufort, "Introduction: Revolution in Military Epistemology," 7–9. Anna Danielsson argues that the "reflexive turn in military epistemology" is a more radical form of critique, as it "addresses possible constitutive effects of knowledge." I do not make a distinction here between these two forms of military design thinking but, rather, use the whole movement as such to reflect on the conjoining of computational trial and error and the experimental ethos of design thinking. See Anna Danielsson, "Knowledge in and of Military Operations: Enriching the Reflexive Gaze in Critical Research on the Military," *Critical Military Studies* 8, no. 3 (2022): 315–333.

126. Graicer, "Self Disruption," 22.

127. Eyal Weizman, *Hollow Land: Israel's Architecture of Occupation* (London: Verso, 2012)

128. See, for instance, John R. Boyd and Grant T. Hammond, eds., *A Discourse on Winning and Losing* (Maxwell Air Force Base, AL: Air University Press, 2018).

129. John H. Holland, *Hidden Order: How Adaptation Builds Complexity* (Reading, MA: Addison-Wesley, 1995).

130. Grant Martin, "Tell Me How to Do This Thing Called Design!: Practical Application of Complexity Theory to Military Operations," *Small Wars Journal* (2011): 4. Emphasis added. https://smallwarsjournal.com/blog/journal/docs-temp/729-martin.pdf.

131. Hito Steyerl, *Duty Free Art: Art in the Age of Planetary Civil War* (London: Verso, 2017), 15.

132. Graicer, "Self Disruption," 32. Emphasis added. This is not unlike "flexible capitalism," where disruption is seen as central to stimulating creativity and innovation, with an emphasis on improvisation over planning. See Benjamin H. Snyder, *The Disrupted Workplace: Time and the Moral Order of Flexible Capitalism* (Oxford, UK: Oxford University Press, 2016).

133. Claudia Aradau, "The Signature of Security: Big data, Anticipation, Surveillance," *Radical Philosophy* 191, May/June (2015): 27.

134. Graicer, "Self Disruption," 28.

135. Tarak Barkawi and Shane Brighton, "Powers of War: Fighting, Knowledge, and Critique," *International Political Sociology* 5, no. 2 (2011), 136.

136. Vivienne Jabri, *War and the Transformation of Global Politics* (London: Palgrave, 2007), 30.

137. On this tension, see Alex Ryan, "A Personal Reflection on Introducing Design to the U.S. Army," Medium, November 5, 2016, https://medium.com/the-overlap/a-personal-reflection-on-introducing-design-to-the-u-s-army-3f8bd76adcb2. In the U.S. Department of the Army, *Army Design Methodology*, ATP 5-0.1 (2015), chapter 5, several traditional planning concepts are included that point toward an incorporation of design into traditional planning and not design as disruption. The tension between the purists and the traditionals was also visible at the yearly military design conference, Innovation Methodologies for Defence Challenges (IMDC) 2019, held at Lancaster University, United Kingdom, on February 26–28, 2019, which I attended.

138. Öberg, "Warfare as Design," 7.

139. See, for instance, U.S. Air Force, *3-60 Targeting, Annex Dynamic Targeting*, (Montgomery, AL: Curtis E. LeMay Center, 2017): 42.

140. Charles Black et al., "U.S. Special Operations Command's Future, by Design," *Joint Force Quarterly* 90, 3rd Quarter (2018): 45.

141. Mitch Ferry, "F3EA—A Targeting Paradigm for Contemporary Warfare," *Australian Army Journal* 10, no. 1 (2013): 58.

142. Massumi, *Ontopower*, 71.

143. Massumi, "Potential Politics and the Primacy of Preemption."

144. Massumi, *Ontopower*, 15.

145. Eyal Weizman, "Thanato-Tactics," in *Beyond Biopolitics: Essays on the Governance of Life and Death*, ed. Patricia Ticineto Clough and Craig Willse (Durham, NC: Duke University Press, 2011), 192.

146. On "nudging" in relation to big data analytics and decision-making see, for instance, Karen Yeung, "'Hypernudge': Big Data as a Mode of Regulation by Design," *Information, Communication & Society* 20, no. 1 (2017), 118–136.

147. On the usage of *nudge* in cybernetics and theories of transformation and the science of change, see James Wilk, "Mind, Nature and the Emerging Science of Change: An Introduction to Metamorphology," in *Metadebates on Science. Einstein*

Meets Magritte: An Interdisciplinary Reflection on Science, Nature, Art, Human Action and Society, ed. G. C. Cornelis, S. Smets, and J. P. Van Bendegem (Dordrecht, The Netherlands: Springer, 1999).

148. This is what military design thinker Orit Gal has termed "social acupuncture." Orit Gal, "Social Acupuncture: Disrupting the Economies of Conflict," paper presented at the Innovation Methodologies for Defence Challenges (IMDC) 2019, Lancaster University, United Kingdom, February 28, 2019).

149. Orit Gal, "Social Acupuncture—An Introduction." https://www.socialacupuncture .co.uk/building-blocks/introduction.

150. In order to rethink the human/machine binary Hayles proposes the distinction between cognizers and noncognizers that foregrounds cognition as the primary analytical category. According to her, humans, animals, and some technical systems, particularly computers, have the ability to process information and generate choice, making them cognizers. For Hayles, it is the ability to process information and generate choice that sets cognizers apart from noncognizers. N. Katherine Hayles, *Unthought: The Power of the Cognitive Nonconscious* (Chicago: University of Chicago Press, 2017), 30–32.

151. Orit Halpern and Gökçe Günel, "Demoing unto Death: Smart Cities, Environment, and Preemptive Hope," *Fibreculture Journal* 29, no. 215 (2017): 2.

152. Michel Callon and John Law, "On Qualculation, Agency, and Otherness," *Environment and Planning D: Society and Space* 23, no. 5 (2005): 717–733.

153. Michiel de Lange, "From Real-Time City to Asynchronicity: Exploring the Real-Time Smart City Dashboard," in *Time for Mapping*, ed. Sybille Lammes et al. (Manchester, UK: Manchester University Press, 2018), 245.

154. Foucault, *Security, Territory, Population*, 132.

155. Rey Chow, *The Age of the World Target: Self-Referentiality in War, Theory, and Comparative Work* (Durham, NC: Duke University Press, 2006), 42.

156. J. Lezaun, N. Marres, and M. Tironi, "Experiments in Participation," in *Handbook of Science and Technology Studies*, ed. U. Felt et al. (Cambridge, MA: MIT Press, 2017).

157. Chandler, *Ontopolitics in the Anthropocene*, 170. Here, Chandler draws on Donna Haraway, *Staying with the Trouble: Making Kin in the Chthulucene* (Durham, NC: Duke University Press, 2016).

158. Chandler, *Ontopolitics in the Anthropocene*, 171.

159. Gregory, "The Everywhere War."

160. Lindsay, "Target Practice."; Nordin and Öberg, "Targeting the Ontology of War."

161. Weber, *Targets of Opportunity*, 105.

162. Simon Glezos, *The Politics of Speed: Capitalism, the State, and War in an Accelerating World* (London: Routledge, 2011).

Chapter 7

1. Jutta Weber, "Keep Adding. On Kill Lists, Drone Warfare and the Politics of Data-bases," *Environment and Planning D: Society and Space* 34, no. 1 (2016): 107–125; Kevin McSorley, "Archives of Enmity and Martial Epistemology," in *(W)Archives: Archival Imaginaries, War, and Contemporary Art*, ed. Daniela Agostinho et al. (Berlin: Sternberg Press, 2021).

2. George E. P. Box, "Robustness in the Strategy of Scientific Model Building," in *Robustness in Statistics*, ed. Robert L. Launer and Graham N. Wilkinson (New York: Academic Press, 1979)

3. Box, "Robustness in the Strategy."

4. Leo Breiman, "Statistical Modeling: The Two Cultures," *Statistical Science* 16, no. 3 (2001): 199–231.

5. Louise Amoore, *Cloud Ethics: Algorithms and the Attributes of Ourselves and Others* (Durham, NC: Duke University Press, 2020), 11.

6. Mike Ananny and Kate Crawford, "Seeing without Knowing: Limitations of the Transparency Ideal and Its Application to Algorithmic Accountability," *New Media & Society* 20, no. 3 (2018): 973–989; Jenna Burrell, "How the Machine 'Thinks': Understanding Opacity in Machine Learning Algorithms," *Big Data & Society* 3, no. 1 (2016): 1–12.

7. Paul N. Edwards, *The Closed World: Computers and the Politics of Discourse in Cold War America* (Cambridge, MA: MIT Press, 1996).

8. Jacquelyn Karpovich, "NGA Director Provides Vision for Next Phase of Intel at GEOINT Symposium," *Pathfinder* 12, no. 2 (2014).

9. Geoffrey C. Bowker and Susan Leigh Star, *Sorting Things Out: Classification and Its Consequences* (Cambridge, MA: MIT Press, 1999).

10. Katherine Hayles and Tony Sampson, "Unthought Meets the Assemblage Brain: A Dialogue Between Katherine Hayles and Tony Sampson," *Capacious: Journal for Emerging Affect Inquiry* 1, no. 2 (2018): 60–84.

11. Donald A. MacKenzie, *Inventing Accuracy: A Historical Sociology of Nuclear Missile Guidance* (Cambridge, MA: MIT Press, 1993). See also Lucy Suchman, "Algorithmic Warfare and the Reinvention of Accuracy," *Critical Studies on Security* 8, no. 2 (2020): 175–187.

12. Nishawn S. Smagh, "Joint All-Domain Command and Control (JADC2)" (Washington, DC: Congressional Research Service [CRS], 2021).

13. David A. Deptula et al., *Restoring America's Military Competitiveness: Mosaic Warfare* (Arlington, VA: Mitchell Institute for Aerospace Studies, September 2019); DARPA, *DARPA Tiles Together a Vision of Mosaic Warfare: Banking on Cost-Effective Complexity to Overwhelm Adversaries*. https://www.darpa.mil/work-with-us/darpa-tiles-together-a-vision-of-mosiac-warfare.

14. Rey Chow, *The Age of the World Target: Self-Referentiality in War, Theory, and Comparative Work* (Durham, NC: Duke University Press, 2006), 42.

Index

ABI (activity-based intelligence), 23,
  175–183, 215
  as algorhythmics, 188–190
  anomaly detection, 177–178, 180–183,
    185–186
  defined, 176
  four pillars of, 177
Abrahams, Edward, 191
Abstraction, 29, 193
Actor network theory (ANT), 137
Adams, Robert H., 73–74
Afghanistan, 1, 104, 129–130, 146
Agency, 38–39, 150, 198, 217, 219–220
  dances of, 69, 76–77
Ahern, Thomas L., 94, 98, 114
Ahmed, Sara, 34
Airplane, 22, 25, 47–50
  aviation, 47–49
Air power, 47–50
  aerial gaze, 32, 47–48
  aerial warfare, 32
  air battle model, 73–75
  air targets, 32, 49, 55, 61, 63, 67–68,
    75–77
  and bombing, 49–50
  and calculus of destruction, 72–73
  colonial air policing, 51–52
  crucial vulnerabilities targeted with, 64
  and enemy, 47–48, 52–54
  in Iraq, 81
  strategic bombing, 49–55, 58–60
Akrich, Madeline, 71
Alexander, Keith, 149–151
Algorhythmics, 188–190, 212
algorhythmic battlespace, 189–190
  algorhythmic governance, 189
  algorhythmic modeling culture,
    196–199
  algorhythmic targeting, 212, 215
  and humans, 220
  and war, 216

Algorithms, 97, 101, 120, 135, 153, 195–198,
  206, 214–217, 219
  algorithmic identity, 156, 184
  algorithmic sieves, 154–157, 172
  beyond doubt, 198–199
  and humans, 214, 217
  interpretability lost, 196
  and knowledge, 216
  and learning, 215
  and population control, 189
  Random Forest, 35, 154–156, 196
  and target production, 180
al-Qaeda, 1
All-Domain Command and Control
  (JADC2), 220
Amoore, Louise, 181, 195, 198–199
Anomaly detection, 177–178, 180–183,
  185–186, 196, 198–199
Aradau, Claudia, 156, 178, 181, 185, 189, 201
Arnold, H. H., 54
Ars Technica, 155
Artificial intelligence (AI), 70, 86, 195, 208,
  214, 216, 219–221
Atomic bomb, 59–62, 72, 77, 86
  calculus of targeting changed by, 72, 77
  nuclear weapons, 63, 72–73, 76–79, 81,
    86, 221
ATP 3-60 (targeting doctrine), 132,
  140–141, 144, 158
Automation, 219–220
  of abstractions, 193
  and creative disruption, 201
  dream of, 10, 70, 125, 174, 192

Bakos, Nada, 136, 161–162
Balibar, Etienne, 14
Barad, Karen, 13, 194, 221
Battlefield, 47
  command and control battlefield man-
    agement, 83–84
  deconstruction of, 84

Battlefield (cont.)
destruction of dynamic targeting on, 82
electronic, 123–127
of the future, 123, 125
and information, 84
order of battle, 87, 91, 116, 133
supplanted by battlespace, 84
targeting defines, 165–167
world rearranged into, 48, 59–60
Battlespace
algorhythmic, 189–190, 205
battlefield supplanted by, 84
and information, 83–84
production of, 32–33
targeting as application of force on elements in, 26
threat network analysis expands notion of, 138
Becoming-with, 193–194, 202, 204–205, 212
Belcher, Oliver, 33, 91–92, 100–01
Biddle, Tami Davis, 53
Big data, 70, 99, 102, 116, 135, 149–153, 191, 195, 197, 201
bin-Laden, Osama, 131
Blanke, Tobias, 156, 178, 181, 185, 189
Bombing, 32
aerial bombardment in Desert Storm, 81
atomic bomb, 59–62, 72, 77, 86
blind-bombing method, 121
British study of German bombing, 55
civilians demoralized by, 52–53, 57–59
displaced and homeless from indiscriminate bombing in South Vietnam, 90
Igloo White rendered bombing an abstract technical exercise, 121, 123
kinds of, 49–50
precision bombing, 51–52, 57, 59–61, 81
Princes Risborough bombing study, 55
shock and awe bombing campaign in Iraq, 83
smart bombs, 81
strategic bombing, 49–55, 58–60
unprecedented destruction of civilian areas, 58–59
US Bombing Survey, 49–51
US Strategic Bombing Survey, 58–59
Bombing Encyclopedia of the World (BE), 22, 61–64, 73, 76–78, 80, 90, 102, 149, 177, 187, 191, 211
data collection for, 65–67
limited utility for war in Vietnam, 87
Bousquet, Antoine, 14, 84

Boyd, John R., 84, 161–164, 200
Bratton, Benjamin, 190
Breiman, Leo, 101, 196, 198, 211
Brigham, Erwin R., 99–100
British Ministry of Economic Warfare (MEW), 57
Bureaucracy, 76–77, 102, 108
Bush, George W., 81

Capturing, 158–159
Carver, George, 104–105
Cetina, Knorr, 102, 165
Chamayou, Grégoire, 41, 143, 151, 166, 180–185, 187, 189, 192
Chandler, David, 207
Child, Elliot, 65
China, 3, 5, 86, 117, 220
Choice, 217
CIA (Central Intelligence Agency), 1, 91–92, 97, 102–104, 106, 113, 116, 136, 161–162, 168, 169
Foreign Industrial Register, 66
Hamlet Evaluation Worksheet, 94
Civilians. *See also* Humans; People
aid in bombing study, 55
captured and killed by RTRG, 172
demoralized by aerial bombardment, 52–53, 57–59
destroying civilians cripples war effort, 60
enemy hiding among, 140
horrific effects of aerial bombardment in Desert Storm, 81
and military, 12, 19, 29, 54, 115, 121, 146–147
strategic bombing shifts carnage to urban centers, 50, 54
unprecedented destruction of, 58–59
warfare as slaughter of, 48, 52, 167
Clapper, James L., 169
Clark, Robert, *Intelligence Analysis: A Target-Centric Approach*, 137
Clausewitz, Carl von, 26, 53, 83, 88, 208
Clinard, Outten J., 67–69
Cognition, 31, 118, 150, 157
cognitive operations, 33–34
cognitive wandering, 67, 69–70, 93, 95, 113, 150
and computers, 197
and machines, 217
Cold War, 1, 3, 6, 10, 20–22, 51, 60–61, 65, 71, 73, 78, 80, 87, 133, 150, 165, 206

Coletta, Claudio, 189
Collier, Stephen, 63–64, 79–80
Colonial air policing, 51–52
COMINT (communication intelligence), 149
Command and control systems, 83–84, 86, 89–90, 108, 130
  All-Domain Command and Control (JADC2), 220
Communication, 49
Computers, 19, 22–23, 31. *See also* Data; Information; Intelligence
  beginning of computerized blacklist, 110–111
  and Bombing Encyclopedia, 63
  and cognition, 197
  computational thinking, 193
  and decision-making, 217
  and humans, 110–111, 114, 124, 147–148, 150, 198, 217–218
  and Igloo White, 120
  and machines, 193, 198
  to make population legible, 91
  and mechanical objectivity, 126
  regime of computation, 199
  self-referentiality of, 165
  subjective human data into objective output, 65, 92–93, 95–97, 100, 126
  sustain imaginary of war as tech problem, 88
  in Vietnam War, 87–90, 92, 95–97, 100–102, 115–16
  war transformed by, 89
  and world, 198, 211, 215
Concrescence, 194
Configuration, 36–37, 39–40
Consolidated Target Intelligence File (CTIF), 67–69
Contact-chaining, 109, 114–115, 136, 140, 143–145, 153, 154, 157, 169
CORDS (Civil Operations and Revolutionary Development Support), 92, 96–99, 103, 105
Counterinsurgency (COIN), 87, 92, 102, 105, 116, 125, 132–133, 135, 163
Counterterrorism (CT), 103–04, 113, 130, 132–3, 152, 160
Couriers, 153–155
Coward, Martin, 31
Critique, 213–214, 218–219, 221
Croizat, Vic, 114

Cybernetics, 78–79
  and situational awareness, 84
  and systems, 88, 90, 127
Cynegetic power, 165–166

Danielsson, Anna, 32, 39
DARPA (Defense Advanced Research Projects Agency), 174
Daston, Lorraine, 13, 197
Data, 19. *See also* Computers; Hamlet Evaluation System; Information; Intelligence
  big data, 70, 99, 102, 116, 135, 150, 153, 191, 195, 197, 201
  Bombing Encyclopedia, data collection for, 65–67
  centrality of data to war, 31, 151, 164
  "collect it all", 149–151, 158, 170, 174, 182, 191
  Consolidated Target Intelligence File (CTIF), 67–69, 72, 95, 150, 177
  databases, 35–36, 89, 109, 111, 115, 148–149, 211
  datafication, 9, 31, 34–35, 65, 212
  data handling, 67–70, 80
  enemy constructed through data infrastructure, 116
  geotagging of, 177
  heterogenous sources combined, 97, 135, 152–153, 158, 167, 171, 176, 178
  and humans, 212
  immersion in data worlds, 193–194
  and learning, 204, 208, 212
  machine methods needed for handling of, 67–69
  mass collection of, 148–151
  massive amounts of, 67, 135, 149, 171
  and models, 196
  production of, 34–36
  querying ability, 36, 61, 69–70, 85, 111, 113–114, 176, 193, 195
  religious belief in, 101
  and sensors, 34–35
  subjective human data into objective output, 65, 92–93, 95–97, 100, 126
  and targeting, 3–5, 29, 216
  translating world into data and data into world, 34, 70, 211–213
  vast assemblage of technicians for processing of, 149
  and violence, 208
Davis, John, 80

Index   **269**

de Goede, Mareike, 136, 138, 164
Deitchman, Seymour, 123
Deleuze, Gilles, 138, 156, 180
  dividuals, 156, 180, 182, 185
  societies of control, 180, 182–183
Delinking, 132, 138, 153
Delori, Mathias, 17
Deptula, David, 82
Destruction
  bodies locked into system of information and, 125
  bombing fuels unprecedented destruction of civilian areas, 58–59
  destruction had to be assured, 72
  of dynamic targeting on battlefield, 82
  electric grids obliterated, 86
  of enemy, 52, 61, 64, 72–73, 80, 85, 124
  of Iraqi systems, 82
  and learning, 135
  and military targeting, 9, 210
  and perception, 84
  targeting to minimize destruction of nuclear attack, 79
Digital age, 149
Digital networks, 2
Dividuals, 156, 180, 182, 185
Dougherty, Christopher, 152
Douhet, Giulio, *Command of the Air*, 52
Drones, 13, 25, 32, 118, 124, 129–130, 135, 147, 152, 167
  "The Drone Papers", 142, 158
Dynamic targeting, 82

Ecology of operations, 44–46
Edwards, Paul N., 48, 101–102, 165, 215
Electric grids obliterated, 86
End-to-end operations, 20, 40–43, 93, 108, 215, 218
Enemy
and air power, 47–48, 52–54
algorhythmic rendering of, 189, 192, 203, 207
  as anomaly, 186
  as anonymous technical signature, 118
  capacity to wage war, 50, 52–53, 55, 63
  capturing better than killing, 159–160
  cognitive operations result in, 33–34
  constructed through data infrastructure, 116
  control perception of, 83
  destruction of, 52, 61, 64, 72–73, 80, 85, 124

emergence of, 163–164, 188, 209
faceless in Vietnam, 87
fantastical models of, 143–144, 186
five-ring model of, 81
force enemy to emerge, 131
as industrial system, 48
learning about, 116, 130–132, 134–35, 159
military targeting as most effective destruction of, 52
models of, 29–30
network, 129, 131, 133–135, 140, 144, 154, 160
as node in network, 125, 134
operationalization of, 22, 26, 29–30, 34, 37, 60, 77–78, 81, 85, 90, 108, 136, 148–149, 151, 165
perturb to learn from, 163–164
from picture of, to model of, 70
production of, 3, 17–19, 22, 26, 30, 88, 108, 115, 135, 138, 148, 164–165, 172, 180
quantifiability of, 88–90, 96–97, 99–100
shift from fixed targets to hunting individuals, 129–131
as system, 48, 59–60, 64, 72–73, 77–78, 80, 85–86, 90, 133
and targeting, 5, 9, 80
threat network analysis expands scope of, 138
unprecedented destruction of, 58–59
unverified, triggered by sensor, 121
as various bits of evidence, 115
various enemies in WWII, 18
in war on terror, 6
Wiener on, 46
and world, continuum between, 9
Enmity, 17–19, 32, 102
  architecture of, 18
  defined, 175
  epistemic production of, 86
  faces of, 17
  signatures of, 151–153
Epistemology. *See also* Knowledge
  defined, 12
  epistemic operations, 13–14, 32–34, 43, 185
  immaterial labour, 69
  martial, 12, 14, 16, 21, 30, 60, 64, 77, 131, 168, 202, 213
  moving military targeting toward questions of, 33–34
  operations at center of, 42
Ernst, Wolfgang, 187

Experimentation, 205–208
Exploratory science, 197

F2T2EA (find-fix-track-target-engage-assess), 133
F3EAD (find-fix-finish-exploit-assess-disseminate), 23, 129–135, 138–140, 145, 150–153, 158–168, 172, 174, 181, 187, 206
  baseball cards, 139–142, 145, 153
  every target is a potential source of intelligence, 134
  the find step, 139–140, 150
  incorporated into targeting doctrine, 132
  to kill or not to kill, 160
  learn about enemy, 130–132, 134
  as network targeting, 133
  as novel targeting methodology, 130, 134
  perturb targets to learn, 162–165
  speed emphasized, 161–162
Faint, Charles, 158
Feminism, 13
Ferry, Mitch, 131, 133–134
Finlayson, Andrew, 114
Flynn, Michael T., 130, 139, 144, 161
Fogleman, Ronald R., 83
Foucault, Michel, 138, 180
  on history of present, 10–11, 36
  on normalization, 180–182
Friend/enemy, 3, 18, 29, 85, 121, 135, 157, 167, 181
Fuller, J. F. C., 54
Full spectrum dominance, 83–84
Future
  beginnings, 23, 46, 199, 202, 204, 208, 210, 210–213, 219
  operationalization of, 83
  past and, coexist, 188, 205
  of politics, 220–221

Gadamer, Hans-Georg, 29
Galison, Peter, 5, 18–19, 55, 78, 197
Garret, Randy, 174
Gathering, 159–160
Gautier, Dave, 176, 193
GEOCELL, 1–2, 169–170
Geo-chaining, 176
Geoghegan, Bernard, 126
Geographic information systems (GIS), 98

GEOINT (geospatial intelligence), 2, 9, 140, 170, 179, 191, 193
Ghamari-Tabrizi, Sharon, 74
Glock, John, 54
Goodman, Steve, 187
Governance, 180, 189, 195
  politics, 220–221
Graham, Steven, 86
Graicer, Ofra, 200–201
Gregory, Derek, 32, 84, 118, 121, 152, 175
Grisogono, Anne-Marie, 164
Grove, Jairus, 14, 44
Günel, Gökçe, 205

Hacking, Ian, 18–9, 107, 180, 197
Hägertsrand, Torsten, 175
Halpern, Orit, 205
Hamlet Evaluation System (HES), 22, 33, 87–88, 90–102, 126–127, 150, 165, 211
  computer central to, 92–93, 100–101
  design of, 92–93
  Hamlet Evaluation Worksheets (HEW), 93–95, 113, 150
  HES Study, 95–96, 99–100
  to make population legible, 91
  and Phoenix program, compared, 108
  veil of objectivity of, 98
  view of below, 100–101
Hansell, Haywood, 53
Haraway, Donna, 101, 194
Harris, Arthur T., 57
Harris, Michael, 158
Harrison, Fisher, 89
Harwood, Graham, 151
Hayden, Michael, 1–2, 5, 131, 168
Hayles, N. Katherine, 33, 45, 184, 198, 217–218
Heart, Liddle, 50
Heath, James, 170–171
Heyck, Hunter, 22, 48, 51, 70–71, 108–109, 197
High-value individuals (HVI), 139–140
Hippler, Thomas, 49, 51
History
  "archives of a future past", 188
  military targeting historically config-ured, 209–210
  of present, 10–11, 36
  targeting in, 25
Hohn, Uta, 57
Holland, Charlie, 131

Index   **271**

Holmqvist, Caroline, 165
Howell, Alison, 12
Humans. *See also* Civilians; Cognition;
  Intelligence; People; Perception
  activity vs behavior, 145
  agency, 39
  and algorhythmics, 219–220
  and algorithms, 214, 217
  anomaly detection moves beyond
    human recognition, 181
  behavior, 22, 52, 69, 95, 109, 114–115, 126,
    144–145, 149, 151–154, 156, 168, 177,
    181, 184
  bureaucracy so dense no one respon-
    sible for mass killing, 76–77
  cognitive wandering, 67, 69–70, 93, 95,
    113, 150
  and computers, 110–111, 114, 124,
    147–148, 150, 198, 217–218
  and data, 212
  expert judgement, 120
  human-machine-environment configura-
    tion, 19, 46, 78, 93, 102, 113, 119, 124,
    126–127, 220
  HUMINT (human intelligence), 140, 143,
    149, 159, 172, 176
  and interpretation, 193
  machine processing rules out human
    judgement, 69
  and machines, 20, 31, 33, 38, 40, 70–71,
    78–79, 88, 90, 93, 97, 118, 121, 135, 148,
    150, 192, 198, 214, 217–220
  and machines, dances of agency
    between, 76–77
  making up people, 18
  and meaning, 199
  morale, 52–53, 57–59
  and nonhumans, 137
  pattern-of-life analysis of, 144–146
  reconfigured by Igloo White, 124
  as sensors, 93, 95, 114
  subjective human data into objective
    output, 65, 92–93, 95–97, 100, 114,
    126
  targeting not done by humans alone, 63
  transformation of conditions of thinking,
    215
  vision, 31–33
  war as nonhuman affair between sys-
    tems, 80
Hussein, Saddam, 131

IBM, 63–64, 66, 90, 97, 110, 115, 119–120,
  143
Igloo White (operation), 22, 88, 117–127
  bombing as abstract technical exercise
    in, 121, 123
  humans reconfigured in, 124
  and sensors, 117–121, 123–124
Imaginaries, 36–37
Immaterial labour, 69
Immutable mobiles, 70
Incitatory power, 163–164, 201, 204
Indexes, 4, 51, 66–67, 70, 105, 110, 180, 182,
  191, 217
Industry
  Industrial Card File (ICF) project, 66
  industrial web theory, 53, 61, 81
  military-industrial complex, 48,
    50–51, 55
  as target, 49, 52–55
Information. *See also* Computers; Data;
  Intelligence
  algorithmic sieves essential to, 155
  basic classes of, 191
  and battlespace, 83–84
  bodies locked into system of, 125
  coordination of, 107–108
  fight for rather than with, 131, 164
  full spectrum dominance, 83–84
  global information grid (GIG), 83
  and learning, 134
  machinic methods for management of,
    62–64, 67
  mass collection of, 149
  massive amounts of, 67, 93, 176
  and meaning, 185
  and military technical revolution (MTR),
    81–82
  operations for, 131–132
  PHMIS (Phung Hoang Management
    Information System), 109–113
  punch cards, 63–64
  targets as sources of, 158–160
  VCINIIS (VCI Neutralization and Identifi-
    cation Information System), 109–110
  in Vietnam War, 87–90
Infrastructure, 80–81, 85, 105–106
  delinking aims for, 132
  electric grids obliterated, 86
  enemy constructed through data infra-
    structure, 116
  shift away from, 129

Intelligence. *See also* COMINT; Computers; Data; GEOINT; Humans; Information; SIGINT
  ABI (activity-based intelligence), 23, 175–183
  actionable, 136
  and air targeting, 67
  artificial intelligence (AI), 70, 86, 195, 208, 214, 216, 219–221
  "collect it all", 149–151, 158, 170, 174, 182, 191
  Consolidated Target Intelligence File (CTIF), 67–69, 72, 95, 149, 177
  coordination of, 106–07
  for counterinsurgency, 116
  enemy as source of, 163
  full spectrum dominance, 83–84
  Intelligence Data Handling System, 70
  lack of, at start of WWII, 54
  little known about South Vietnamese population, 91
  and operations, 130, 152, 160, 167
  provincial reconstruction units (PRU), 113–114
  and strategic bombing, 65
  strategic intelligence needed for strategic warfare, 53
  targeting intelligence, 61, 70
  targets as sources of, 134, 158–161
  translate environment into actionable, 88
The Intercept, 142, 158
Interpretation, 70, 95, 185–186, 197, 214–215
  accuracy over, 199
  algorhithms lose, 196
  and humans, 193
Interrogation, 104–105, 159
Iraq, 3, 5, 81, 104, 129–131, 159–161, 170–171
  electric grids obliterated, 86
  Most Wanted Identification Playing Cards, 131
  shock and awe bombing campaign in, 83
Israel, 138, 163, 199

Jacobsen, Annie, *First Platoon*, 146
Japan, 5, 58–61
Jason Group, 117, 123
Johnson, Kenneth T., 70, 74–76
Joint targeting cycle, 26–27, 42

Kent, Alexander J., 80
Kilcullen, David, 132
Kinsley, Samuel, 187
Kitchin, Rob, 189, 197
Klee, Paul, 182
Knapp, William, 106–107
Knowledge. *See also* Epistemology
  and algorithms, 216
  and Bombing Encyclopedia, 64
  error of, 201
  everything, everywhere, all at once, 191, 203
  geographical knowledge production, 32
  and history of present, 10–11
  as learning, 207
  little known about South Vietnamese population, 91
  and martial, 13, 17
  and military design, 200
  of objects, 13
  operationalization of, 31
  and operations, 42
  production of, 13–14, 33, 42–43
  and war, 21, 101, 221
Known knowns, 5–6, 20, 30, 133
  in matrix of targeting, 7–8
Known unknowns, 6, 20, 22–23, 152, 168, 199
  in matrix of targeting, 7–8
Kockelman, Paul, 155
Komer, Robert William, 92–93, 95, 99, 106

Lakoff, Andrew, 63–64, 79–80
Lamb, Christopher J., 161
Large language models (LLMs), 198
Latour, Bruno, 38, 70, 137
Learning
  and algorithms, 215
  and data, 204, 208, 212
  about enemy, 116, 130–132, 134–135, 159
  through error, 201
  knowledge as, 207
  and military, 200–201, 205
  perturb targets to learn, 162–165
  and targeting, 23, 31, 131, 206
Leavitt, Robert W., 78
Lefebvre, Henry, 175, 182
Liberalism, 183
Lindsay, John, 162
Literacy, 41
Ljungkvist, Kristin, 39

Index   **273**

Long, Letitia, 5–7, 176, 193
Lowe, James T., 62, 67

Machine learning, 135, 153, 156–157,
    195–198, 212, 214, 216
  and anomaly detection, 181
  beyond doubt, 198–199
  core elements of, 155
  questions about, 217–218
  Random Forest, 35, 154–156, 196
Machines
  and cognition, 217
  and computers, 193, 198
  and data handling, 67–70
  human-machine-environment configura-
    tion, 19, 46, 78, 93, 102, 113, 119, 124,
    126–127, 220
  and humans, 20, 31, 33, 38, 40, 70–71,
    78–79, 88, 90, 93, 97, 118, 121, 135, 148,
    150, 192, 198, 214, 217–220
  and humans, dances of agency between,
    76–77
  and information management, 62–63
  machine processing rules out human
    judgement, 69
  machinic haze, 215
  questions for, 217–218
  and reality, 198
  and targeting, 72, 77
  and uncertainty, 203
  war transferred from people to,
    124
MacKenzie, Donald, 30
Managerial approach to war, 87–90, 92,
    125
Manhunting, 1, 3, 106, 113, 116, 130, 134,
    152, 161, 210, 220
  war as hunting, 166
Manipulable mobiles, 71
Map of the World (MoW), 9, 190–194, 205,
    211–213, 215–216, 220
Martial
  defined, 12
  empiricism, 14
  epistemology, 12, 14, 16, 21, 30, 60, 64, 77,
    131, 168, 202, 213
  gaze, 33, 84
  and knowledge, 13, 17
Massumi, Brian, 30, 45, 83, 162, 165, 203
  incitatory power, 163–164, 201, 204
Matrix of targeting, 6–8, 20–21, 190
McChrystal, Stanley, 129–130, 161

McDonald, General, 59
McNamara line, 121
McNamara, Robert, 89, 104, 109, 117, 125
McSorley, Kevin, 69
Meaning, 25–26, 29, 31, 33, 36, 38, 45, 150,
    214–215, 217, 219
  and humans, 199
  and information, 185
Mégret, Frédéric, 84
Mellinger, Phillip S., 47
Merton, Robert, 30
Military
  and civilians, 12, 19, 29, 54, 115, 121,
    146–147
  design, 195, 199–204, 206, 208, 213, 219
  imaginaries, 36–37
  industrial complex, 48, 50–51, 55
  and learning, 200–201, 205
  military technical revolution (MTR),
    81–82
  network-centric warfare (NCW) as fun-
    damental shift in, 83
  revolution in military affairs (RMA), 200,
    220
  and war, 15–16
Military targeting, 3–5. *See also* Targeting
  emergence of, 22
  as an end in itself, 23, 165
  great paths of destruction in wake of, 9
  historically configured, 209–210
  beyond individuals, 192
  is different, 183–185
  and knowledge, 207
  as most effective way to destroy modern
    enemy, 52
  and questions of epistemology, 33–34
  rarely focused on, 25
  seeds of, 50
  and threat emergence, 204
  unfathomable number of victims, 210
  and war, 210
  world according to, 211, 221
Miyazaki, Shintaro, 188–189
Mnemonic control, 187
Mnemotechnics, 187
Models, 29–31
  air battle model, 73–75
  algorhythmic modeling culture,
    196–199
  "all models are wrong", 212
  and data, 196
  five-ring model of enemy, 81

imaginary world of model as policy action in real hot war, 74
operational models, 70–71
religious belief in, 101
Morale, 52–53, 57–59
Mosaic Warfare, 220
Mullins, John, 106
Multi-INT, 149, 158–159, 170, 176, 191
Munsing, Evan, 161

Nagl, John, 132, 135, 160
Nash, Henry, 76–77
Naveh, Simon, 163
Networks, 115
actor network theory (ANT), 137
digital networks, 2
enemy network, 129, 131, 133–135, 140, 144, 154, 160
linking seeds causes networks to emerge, 143–144
network-centric warfare (NCW), 83, 124, 220
network thinking, 45, 108–109, 114
neural networks, 35, 196–198, 214, 219
social network analysis (SNA), 109, 135–138
vs systems, 88, 108–109
target critical nodes in, 132, 134, 136–137
threat network, 137, 139, 145, 150, 158
New world order, 81
NGA (National Geospatial-Intelligence Agency), 1–2, 169–170, 190
Future State Vision, 190
Map of the World, 9, 190–194, 205, 211–213, 215–216, 220
Niva, Steve, 129
Nixon, Richard, 125
Nordin, Astrid, 16, 39, 84, 165
Normal, 175, 177, 180
normalization, 180–182
NSA (National Security Agency), 1–2, 171, 173
"collect it all" mantra, 149–151, 158, 170, 174, 182, 191
Cryptologic Support Group (CSG), 169–170
SKYNET, 154–158
Nuclear war, 22, 60, 62, 72–74, 77, 80, 87, 100, 127
mutually assured destruction (MAD), 80

Nuclear weapons, 63, 72–73, 76–79, 81, 86, 221
atomic bomb, 59–62, 72, 77, 86

Öberg, Dan, 15–16, 39, 84, 164, 202
OODA (observe-orient-decide-act) loop, 84, 161, 163
Operation Desert Storm, 81
Operations
analysis, 55
defined, 41–42
ecology of, 44–46
effects-based operations, 82–83
epistemic operations, 13–14, 32–34, 43, 185
for information, 131–132
and intelligence, 130, 152, 160, 167
operational environments, 32
operational models, 70–71
operative logics, 30
Other, 102, 220
enemy as, 17–18, 32, 54

Pacification, 91–97, 99–100, 105
Packer, Jeremy, 69
Palantir, 143, 147
Pang, Guansong, 196
Parisi, Luciana, 187, 194, 203
Pasquinelli, Matteo, 155, 183
Pattern-of-life (PoL) analysis, 135, 144–146, 152–153
vs contact chaining, 145
making up the wrong people, 144, 146–148
People. *See also* Civilians; Cognition; Humans; Perception
boundary conditions for governing, 180
devastating effects of violence on, 43
displaced and homeless from indiscriminate bombing in South Vietnam, 90
enemy hiding among, 140
little known about South Vietnamese population, 91
making up, 18, 144, 146–148, 153, 167
mass interrogation of, 65
morale of, 52–53, 57–59
Phoenix program swept up innocent people, 114–116
quantification of war neglects, 100
targeted societies, 180, 182–183, 192
targeting bodies and behavior rather than persons, 115

People (cont.)
  unfathomable number of victims of military targeting, 210
  war as terrible devastation of, 23
  war transferred from people to machines, 124
Perception, 31–33, 118, 125
  control enemy's, 83
  and destruction, 84
Phoenix program/Phung Hoang, 22, 88, 98, 102–116, 126–127, 149, 211
  advisors' handbook, 107–108
  five programs of, 103–104
  and Hamlet Evaluation System, compared, 108
  innocent people swept up in, 114–116
  PHMIS (Phung Hoang Management Information System), 109–113
  spread of, 116, 132
Politics, 26, 220–221
  governance, 180, 189, 195
Population. *See* Civilians; People
Portal, Charles, 57
Precision bombing, 51–52, 57, 59–61, 81
Presentism, 188
Princes Risborough bombing study, 55
Processes, vs systems, 45
Pugliese, Joseph, 146, 152–153
Puig de la Bellacasa, Maria, 45

Quantification, 88–90, 96–97, 99–100

Raetz, William, 176, 178
Raley, Rita, 195
Random Forest, 35, 154–156, 196
Reality, 29, 32, 98–99, 174, 198, 215
Reeves, Joshua, 69
Relationality, 29, 137, 151, 184
Resor, Stanley, 114
Revolution in military affairs (RMA), 200, 220
Rheinberger, Hans-Jörg, 13
Rosenblueth, Arturo, 29–30, 197
Rottman, George L., 87
RTRG (real-time regional gateway), 169–174
  capture/kill data of, 172
  not engineered for accuracy, 172, 174
  software applications in, 172
  staggering volume of data from, 171
Rumsfeld, Donald, 6, 203

Russia, 66, 86
Ryan, Alex, 164

Said, Edward, 17
Schenecker, Parker, 174
Science and Technology Studies (STS), 13, 43, 206
Scientific rationality, 197
Scott, James, 33
Secrecy, 43
Security, 181–182, 195
Seeds, 143
Semi-Automatic Ground Environment (SAGE), 89–90
Sensitivity analysis, 74
Sensors, 34–35, 80–84, 99, 135
  and blind-bombing method, 121
  growing family of, 149
  humans as, 93, 95, 114
  Igloo White all about, 117–121, 123–124
  "sensors saved U.S. lives", 123
  sensor-to-shooter system, 118
  targeting requires, 216
  unverified enemy triggered by sensor, 121
  war reduced to issue of, 88
  world as, 189
Shah, Nish, 14
SHARKFINN, 169, 172
SIDtoday, 174
SIGINT (signals intelligence), 2, 140, 143, 149, 158–159, 170–172, 176–177
  "coin of realm" in war on terror, 149
Signatures
  signature of terrorist, 152–153
  signature recognition, 151–153
  from signatures to anomalies, 180, 183
Situatedness, 4, 13, 29–32, 38, 41–43, 109, 114, 116, 134, 184–185, 194, 212, 216
Situational awareness, 84
SKYNET (NSA), 154–158
Smart bombs, 81
Smart cities, 189
SNA (social network analysis), 109, 135–38, 178
  replaced by TNA (threat network analysis), 137–138
Snowden, Edward, 8, 149, 152, 154
Societies of control, 180, 182–183

Soviet Union (USSR), 1, 5–6, 60, 62–63, 65–66, 72–73, 75, 77, 79–81, 86–87
  Red Atlas, 80
Star, Leigh, 45
Statistics, 66, 91–92, 97–101, 110–111, 114–115, 119, 155–156, 177–178, 180–182, 212
  two cultures of, 196
Steigler, Bernd, 187
Steyerl, Hito, 153, 201
Strategic bombing, 49–55, 58–60
  atomic bomb, 60–61
  sociotechnical conditions of possibility for, 66
  and useful intelligence, 65
Strategic vulnerabilities, 79–81
Strategy, 200–201
Suchman, Lucy, 36, 38, 40, 125, 194, 218
Summers, Harry G., 125
Surveillance, 149, 151
Systems
  age of, 22, 48, 51, 59, 64
  analysis, 51, 78–79, 81, 88–90, 97, 109, 127, 136
  command and control systems, 83–84, 86, 89–90
  and cybernetics, 88, 90, 127
  enemy as, 48, 59–60, 64, 72–73, 77–78, 80, 85–86, 90, 133
  five-ring model of enemy systems, 81
  vs networks, 108–109
  vs processes, 45
  and strategic bombing, 51–52, 59–60
  from systems to networks, 88
  and target selection, 54–55, 57, 59, 72
  vital systems, 79–80
  war as nonhuman affair between systems, 80
  world as, 48, 85

Targeted societies, 180, 182–183, 192
Targeting. *See also* Military targeting
  algorhythmic, 212, 215
  as application of force on elements in battlespace, 26
  atomic bomb changed calculus of, 72, 77
  ATP 3-60 (targeting doctrine), 132, 140–141, 144, 158
  bodies rather than persons, 115
  contemporary function of, 26
  cyclical nature of, 165

dynamic targeting, 82
and enemy, 5, 9, 80
everything, everywhere, all at once, 3, 23, 151, 187, 203, 208, 221
F3EAD as novel methodology for, 130
F3EAD incorporated into targeting doctrine, 132
global imperium of, 84
in history, 25
imaginaries fundamental to, 37
is operational, 30
joint targeting cycle, 26–27, 42
as key to survival in world of nuclear superpowers, 79
lack of proper targeting analysis in WWII, 54–55
and learning, 23, 31, 131, 206
and machines, 72, 77
matrix of, 6–8, 20–21, 190
not done by humans alone, 63
representations of, 25
seeds of, 50
target analysis, 55–56, 58
targeting intelligence, 61, 70
as technoscientific undertaking, 64
as transhistorical transformation, 21
unknown becomes targeted, 183
violence of, 43, 159–160
and war, 3–5, 80, 131, 161, 164–165, 167, 220
war as endless targeting practice, 164–165
of world, 3–5, 10, 26, 31, 64, 85, 205–208, 211, 213, 221
Targets. *See also* Bombing Encyclopedia
  air targets, 32, 49, 55, 61, 63, 67–68, 75–77
  algorithmic production of, 180
  always partial, 186
  anything that moves, 83
  Bombing Encyclopedia for target recommendations, 61
  Consolidated Target Intelligence File (CTIF), 67–69
  endless production of, 150
  expanded notion of, 82
  importance of, in relation to other targets, 26, 28
  industrial centers as, 49, 52–55
  as information sources, 158–160
  perturb targets to learn, 162–165, 203

Index   277

Targets (cont.)
production of, 32, 139, 150, 164
shift from fixed targets to hunting individuals, 129–131
subjective human data into objective targets, 65
target analysis, 55–56, 58
target selection, 54–55, 57, 59, 61, 72
threat network analysis expands scope of, 138
worm targets, 121–122
Task Force Alpha (TFA), 118, 120
Technocracy, 126–127
Technogenesis, 186
Technopolitics, 219–220
Techno-rationality, 197
Technoscience, 195, 197, 213, 219
Technowar, 89
Terrorist, identification of, 146, 151–155, 168
Thayer, Thomas C., *A Systems Analysis View of the Vietnam War*, 97
Thomas, Jim, 152
Threats, 203–04
TNA (threat network analysis), 137–138, 178
Total war, 48, 54, 59
"Track 'em and whack 'em", 1, 130, 170
Trench warfare, 50
Trial and error, 23

UK Government Communications Head Quarters (GCHQ), 143, 170
Ulman, Harlan, 83
Unknown knowns, 6, 20, 22–23, 133, 151, 168
in matrix of targeting, 7–8
Unknown unknowns, 6, 20, 23, 156, 168, 174, 182, 199, 203
choice method to uncover, 175
in matrix of targeting, 7–8
US Air Force
Directorate of Targets, 67
F2TEA (find-fix-track-target-engage-assess), 133
in Igloo White, 117
Intelligence Data Handling System, 70
Operation Desert Storm, 81
Project Wringer, 65
Strategic Vulnerability Branch (SVB), 61–62, 64
Target Analysis Division, 63

US Army
Air Corps Tactical School (ACTS), 5, 52–54, 81
Air Force (AAF), 55
D3A (decide-detect-deliver-assess), 133
F3EAD incorporated into targeting doctrine, 132
Industrial Card File (ICF) project, 66
Remotely Monitored Battlefield Sensor System (REMBASS), 124
Special Documents Section (SDS), 66
targeted killing as American way of war, 168
US Bombing Survey, 49–51
US Central Command (CENTCOM), 174
US Committee of Operations Analysts (COA), 55, 62
US Defense Department, 76, 89
US Enemy Objective Unit (EOU), 55–57, 62
US joint targeting cycle, 26–27, 42
US Joint Targeting Group (JTG), 55, 62
US military
Assistance Command Vietnam (MAVC), 97, 105
Bombing Encyclopedia of the World (BE), 61–64
computational view prevalent within, 100
data central to, 31
drowning in data, 99
FM 30-5, *Combat Intelligence* (field manual), 116–117
full spectrum dominance, 83–84
ill-prepared for war in Vietnam, 87
Joint Special Operations Command (JSOC), 129, 160
little known about South Vietnamese population, 91
matrix of targeting, 6–7
network-centric warfare (NCW) transformation of, 82–83
Operation Desert Storm, 81
Operation Igloo White, 117–125
operations based on information feedback loops, 90
shift from fixed targets to hunting individuals, 129–131
Special Operations Command (SOCOM), 131, 137
Special Operations Forces (SOF), 130, 149, 158–162, 166–167, 172

targeted killing as American way of war, 168

unblinking eye, 144

war reorganized by, 1–3

US National Security Agency (NSA), 143

US Navy, 124

US Strategic Air Command (SAC), 60, 64, 72–73

US Strategic Bombing Survey (USSBS), 58–59

Valentine, Douglas, 109

Vietnam/War, 22–23, 33
  Census-Grievance program, 104–105, 114
  Combined Intelligence Center Vietnam (CICV), 103, 105
  computers and information processing in, 87–90, 92, 95–97, 100–102, 115–116
  disaster of, 114–116
  as experimental laboratory, 87–88
  as first technowar, 89
  Hamlet Evaluation System (HES), 90–102, 126–127
  little known about South Vietnamese population, 91
  pacification effort in, 91–97, 99–100, 105
  PHMIS (Phung Hoang Management Information System), 109–113
  Phoenix program/Phung Hoang, 22, 88, 98, 102–116, 126–127
  provincial reconstruction units (PRU), 113–114
  US National Archives on, 97
  VCINIIS (VCI Neutralization and Identification Information System), 109–110
  "when will we win?", 125

Vision, 31–33

Vital systems, 79–80

Voelz, Glen, 130

Wade, James, 83

Walsh, William, 60, 77

Warden, John, 81

Ware, Vron, 17

War on terror, 1, 3, 10, 20, 104, 130, 135, 137, 149, 166–167, 211
  enemy of, 6

War/warfare
  before air power, 47
  and algorhythmics, 216

architecture of enmity as condition for, 18

computers sustain imaginary of war as tech problem, 88

computers transformed, 89

as constitutive of social and political orders, 12, 15

data central to, 31, 151, 164

destroying civilians cripples war effort, 60

dislocated from territory, 165–166

emergent character of, 209

as endless experimentation, 206–208

as endless targeting practice, 164–164

enemy capacity for, 50, 52–53, 55, 63

experimental approach to, 201

as experimental vision of world, 202

horrors of, 210–211

human-machine-environment configurations structure bureaucracy of, 102

as hunting, 166

imaginary world of model as policy action in real hot war, 74

and knowledge, 21, 101, 221

managerial approach to, 87–90, 92, 125

and military, 15–16

military targeting central to, 210

network-centric warfare (NCW), 82–83

as nonhuman affair between systems, 80

politics of, 26

as processing, 16

prosthetic warfare, 84

psychological effects of, 160

quantifiability of, 88–90, 96–97, 99–100

as question, 14–15

questions about, 16, 217–218

rationalization of, 15, 22–23, 26

reduction of, for strategists, 86

reorganized, 1–3

secrecy around, 43

speed emphasized, 161–162

still not datafied, 212

and targeting, 3–5, 80, 131, 161, 164–165, 167, 220

technowar, 89

as terrible devastation, 23, 43, 48, 54

total war, 48, 54, 59

transferred from people to machines, 124

trench warfare, 50

war into warfare, 16, 85, 164, 202

Weber, Jutta, 40, 151, 164, 195, 197

Weizman, Eyal, 137, 163

Westmoreland, William, 123, 125
Whitehead, Alfred North, 193–194
Wiener, Norbert, 18, 29–30, 46, 79, 193, 197
Wing, Jeanette, 193
World. *See also* Map of the World
  abstraction of, 29
  and algorithms, 195–196
  alternatives impossible to envision, 208
  battlefield remakes world, 84
  cannot be governed or secured, 195
  closed world, 5, 10, 14, 64, 79, 84, 101,
    165, 206, 215
  complexity of, 197
  computerized view of, 89–90
  and computers, 198, 211, 215
  digital twin of, 9, 191–192, 221
  and enemy, continuum between, 9
  as hunting ground, 166
  imaginary world of model as policy
    action in real hot war, 74
  Map of the World, 190–194
  monitor entirety of, 151
  new world order, 81
  operationalization of, 22, 29, 34
  operational models of, 70–71
  from picture to model of, 70–71
  production of, 17–18
  as sensor, 189
  and strategy, 201
  as system, 48, 85
  of targeting, 3–5, 10, 26, 31, 64, 85,
    205–208, 211, 213, 221
  targeting is key to survival in world of
    nuclear superpowers, 79
  translating world into data and data into
    world, 34, 70, 211–213
  warfare as experimental vision of, 202
World War I (WWI), 5, 32, 47–50, 52, 54
  US Bombing Survey, 49–51
World War II (WWII), 10, 22, 48, 60–62,
  78–79, 86–87, 89–90, 139, 221
  advanced preparation for future contin-
    gencies not present in, 64
  enemies in, 18
  lack of proper targeting analysis in,
    54–55
  US Strategic Bombing Survey, 58–59
Worm targets, 121–122

Zaidan, Ahmad, 156
Zarqawi, Abu Musab al-, 131, 162
Zawahiri, Ayman al-, 131